FARMING AS A BUSINESS

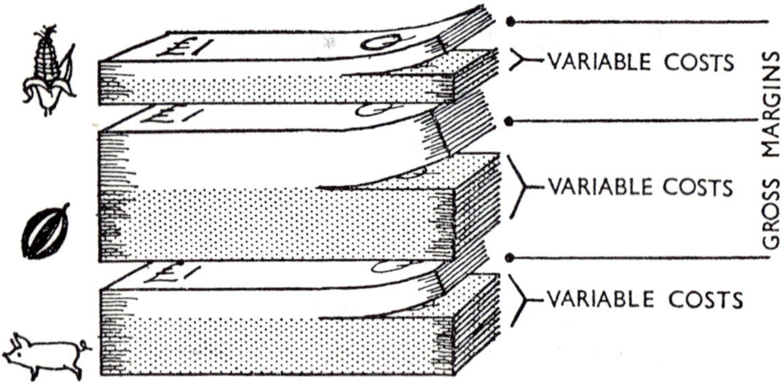

The value of production less variable costs equals the gross margin for each enterprise.

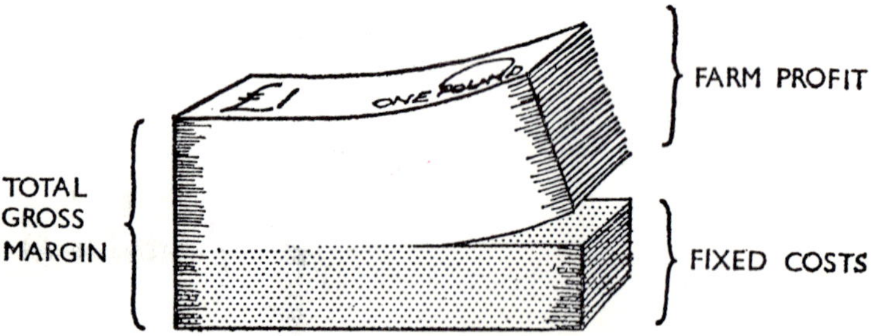

The total gross margin less fixed costs equals the farm profit.

FARMING AS A BUSINESS

BY

M. UPTON

and

Q. B. O. ANTHONIO

Department of Agricultural Economics, University of Ibadan

OXFORD UNIVERSITY PRESS

Oxford University Press, Ely House, London W. 1

GLASGOW NEW YORK TORONTO MELBOURNE WELLINGTON
CAPE TOWN SALISBURY IBADAN NAIROBI DAR ES SALAAM LUSAKA ADDIS ABABA
BOMBAY CALCUTTA MADRAS KARACHI LAHORE DACCA
KUALA LUMPUR SINGAPORE HONG KONG TOKYO

OXFORD TROPICAL HANDBOOKS

A. E. HAARER : *Coffee Growing*
C. R. HARLER : *Tea Manufacture*
C. J. PIGGOTT : *Coconut Growing*
A. H. SAVILE : *Extension in Rural Communities*
M. UPTON & Q. B. O. ANTHONIO : *Farming as a Business*

First published 1965
Reprinted lithographically in Great Britain
(with corrections)
at the Pitman Press, Bath
1970

Preface

Many people consider farming a lowly occupation. Most fathers hope their sons will not be farmers but will get work in the towns. These people do not realize the great importance of farming. It should be remembered that everybody from the Prime Minister down to the beggar on the streets depends upon farmers for his food.

In many tropical countries the farmers and their families not only produce food for the people but goods for export as well, such as cotton, rubber, sugar or cocoa. These are the main exports of many tropical countries. The wealth of such countries depends upon the farmers.

The development of manufacturing industries in tropical countries requires capital and labour. These resources must come from the farming community. Capital is saved out of the sales of farm produce and labour comes from those members of farming families who move to the towns.

Obviously farming is a very worthwhile and important occupation and should not be despised.

Many people who would agree that farming is important would say 'I do not want my son to be a farmer because farmers are poor'. This is not necessarily true. Farmers need not be poor. Many farmers earn more profit than an office messenger or a lorry driver. A farmer who is a good manager should be able to make as much profit from farming as from any other business.

This book does not set out to tell any individual farmer exactly what to do to make high profits. Reading this book will not automatically make a farmer a good manager. But if the advice given is taken, if the farmer starts keeping some records and accounts and if he starts planning his farm and carrying out those plans, he should become a better manager.

Ibadan M.U.
October, 1964 Q.B.O.A.

Acknowledgements

The authors wish to thank all those individuals who helped by reading the early drafts and making helpful suggestions and, in particular, Mr H. T. E. Smith, Senior Lecturer in the Department of Agricultural Organization, University of Ibadan, who gave helpful advice throughout the preparation of the text and diagrams, and Mr Peter Upton who drew all the diagrams.

Note

To avoid any possible confusion which might occur if reference were given to several different currencies, the common decimal currency £0·00 exclusively has been used throughout this book. However, the business principles discussed and the advice the text gives regarding the keeping of accounts, etc., are, of course, applicable in any currency.

Similarly, where reference is made to weights, and measures of capacity, length and area, the units given are SI metric. For those who wish to convert these into other equivalents, a quick-comparison chart is given in the endpapers.

Contents

1 Introduction

The Importance of farm management

A farmer needs to know not only how to cultivate his crops and tend his livestock but also how to manage his farm. Many farms consist of several different sections, each devoted to the production of one kind of crop or livestock. These sections of the farm are known as enterprises. Every farmer must face the management problems of deciding which enterprises to have, how much to produce in each enterprise and what methods to use.

In the past, advisory work among farmers has been almost entirely concentrated upon the technical problems of producing crops and livestock. Now it is realized that improvement in the farmer's technical knowledge should go hand in hand with the development of his managerial ability. Farm management is now an important part of the study of agriculture.

Productive resources

Apart from the skill and knowledge necessary to manage his farm, a farmer also needs a supply of land, labour and capital. Land is needed for crops to grow and livestock to graze, labour is needed to do the work and capital is needed in the form of equipment, buildings, stores, permanent crops and livestock. Land, labour and capital are known as productive resources. The farmer's skill and knowledge as a manager may be considered as another productive resource.

The amount a farmer can produce is limited by his supply of productive resources. Where there are large areas of land available for cultivation, farms are limited by the supply of labour or capital. In other places, where there are large numbers of labourers available for employment, land or capital may limit the amount each farmer can produce.

Most enterprises compete with each other in the use of productive resources. This means that to produce more from one enterprise a farmer must produce less from another enterprise. To be able to produce more cocoa or coffee he must produce fewer food crops. In planning the size of each enterprise the farmer is at the same time planning how to use his resources. The problems of farm management are the problems of how best to use the limited supplies of productive resources.

The farmer's objectives

In deciding how to use his productive resources a farmer must be guided by certain objectives. Many farmers use traditional methods and follow the old established pattern of farming. Traditional methods

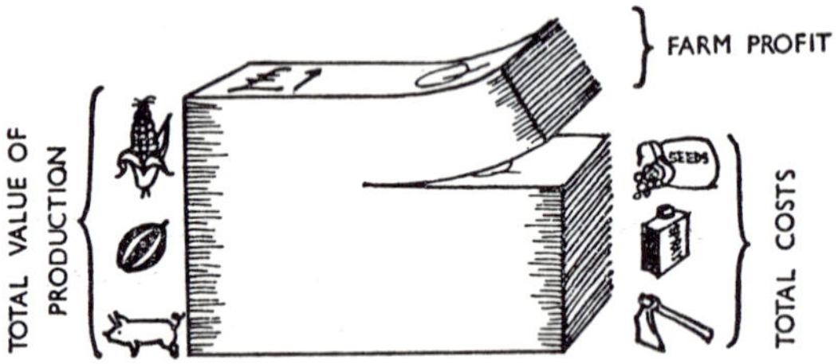

Fig. 1. Total value of production less total costs equals farm profit.

have grown up over a long period and have been well tried. However, the conditions under which farming is carried on are changing fast. To be efficient, farmers must be prepared to learn and use new methods of production.

The aim of some farmers is to produce just enough food for the needs of themselves and their families. This means that they do not sell any produce. Farming to produce just sufficient food for the family is called subsistence farming. Nowadays few farmers are satisfied by subsistence farming. They want money to buy bicycles, radios, watches, expensive clothes and new foods imported from other countries. Money is also needed to pay children's school fees, doctors' bills and government taxes. Practically all farmers now sell some produce and true subsistence farming has ended.

Today the aim of most farmers is to earn profits. Profit is the difference between the value of production and costs (Fig. 1) and represents the farmer's gain. Many crops are grown only to be sold for profit. The tropical climate is especially suitable for certain valuable crops which will not grow elsewhere. People from the colder regions want to buy tea, coffee, cocoa, tobacco, cotton, rubber, fibre crops, groundnuts, coconuts, palm oil and other palm produce, sugar, rice and bananas from the tropical countries. Tropical farmers can grow and sell all these for profit.

The traditional food crops can also be sold for profit. The farmers growing only export crops and the people working in industry and commerce in the towns need food which must be bought. It is quite common for crops such as yams or maize to be grown and sold for profit in the local village markets and in the towns. Even the nomadic herdsmen like the Fulani of West Africa and the Masai of East Africa sell their stock for meat in the towns.

Farming as a business

Farming as a business means farming for profit, not just *some* profit but as much profit as possible. A farmer who uses his productive resources to grow food crops worth £60 at a cost of £10 is making a profit of £50. But if he is a real businessman he should be asking himself whether he is making as much profit as possible. He is not, if the same resources could be used to earn £70 profit growing rubber or some other crop.

This does not mean that the business farmer must sell all his produce. In fact, there are certain advantages in producing food for the family on the farm. It is usually cheaper to grow food than it is to buy it; less time is spent shopping in the market and home-grown food is fresher. However, the business farmer cannot be content with subsistence farming and expects to sell a large part of his produce.

The two sides of business farming

Farm business management is concerned with using the farmer's limited resources to give the highest profits. There are two sides to this problem. The first is the long-term planning of the use of these resources and the other is the day-to-day management. Both are important in getting high profits.

For planning, the farmer must have time to work out his ideas, preferably writing them down and estimating the cost and value of production for each plan. The day-to-day management may involve taking quick decisions, especially if plants or animals become diseased or if there are not enough workers to finish a job in time. Farm records can show the farmer whether his day-to-day management is successful.

The rest of this book deals with ideas which can help farmers in both planning and managing for high profits. Unfortunately no two farms are alike so it is impossible to give a set of rules for profitable farming which all farmers should follow. We can set down basic principles for all farms. But when it comes to deciding just what crops and livestock will be produced, how they will be produced and in what quantity, each farm must be considered separately. Each farmer must find solutions to his own management problems, although he may be able to get help from government advisory officers.

A study of the basic principles of farming as a business and methods of keeping records and preparing farm plans is important to all farmers and farm advisers.

2 Some Basic Principles

The value of farm production

Fig. 1 illustrates the relationship.

PROFIT = TOTAL VALUE OF PRODUCTION LESS TOTAL COSTS.

The total value of production is more than just the money received from sales of produce. Stored crops and food used in the house have been produced, even though they are not sold. The growth of young tree crops and livestock is also a form of production. As a tree or an

Fig. 2. The production of a growing tree or animal is measured by the increase in value.

animal grows it gains in value; a mature oil palm is worth more than a seedling and an adult pig is worth more than a young one. The increase in value of the tree or animal represents the value of what has been produced (Fig. 2). Total production therefore includes produce sold, produce used on the farm and the growth of young trees and livestock.

The value of produce sold is easily measured by the amount of money received for it. The value of unsold produce is more difficult to measure. There are two possible methods. One way is to use the market price; that is, the price for which the produce could have been sold. This is the simplest method of estimating the value of produce used on the farm. It may be possible to estimate the value of animals in the same way. Trees, on the other hand, are not usually bought and sold in the market so another method must be used. The value of growing trees can be taken as the cost of production, which is the total cost of growing the tree up to its present age.

Total value of production, cost and profit must be measured over a fixed period of time. The most convenient period for most purposes is 1 year, although a poultry farmer or a dairy farmer may want to check on his production week by week.

Farm prices

The price a farmer can get for his produce depends upon both supply and demand. Supply is the amount available for sale and demand is the amount people wish to buy. If the supply increases without the demand changing, the price will fall. This is the reason why crop prices fall at harvest time when there is plenty available for sale in the markets. If demand increases without the supply changing the price is likely to rise. Where the people of a country like to celebrate Christmas by eating turkey, the demand and the price for turkeys rises just before Christmas. In Islamic countries where the people celebrate 'Id al-adha' by killing a ram, the demand and price for rams rises just before this festival. Many other examples could be found of supply and demand affecting prices.

If a farmer knows when to expect price changes he can plan to sell when prices are high. By harvesting and selling his crops early, before most other farmers, he might avoid the low prices at harvest time. This means that he must plant early. Alternatively he may store his crops at harvest time to sell or use them later when market prices have risen. Turkey farmers can get higher prices by making sure that most of their birds are ready for sale just before Christmas. Sheep farmers can do the same by producing rams ready for sale at 'Id al-adha'.

These are a few examples of how farmers can arrange to sell when prices are high. By getting higher prices a farmer increases the value of production and gets higher returns. However, early harvest or storage of crops and production of livestock for a special season requires careful management and may involve extra labour and capital. Extra resources are used in order to get the extra returns. Before using these methods of getting extra returns the business farmer should first make sure that he cannot get even higher returns by using the extra resources in some other way.

Quality and price

The price obtained for many farm products depends upon the quality. Women will pay more for rice which has no dirt or stones in it and is free of insects, and this is true for most food crops. Large eggs will sell for a higher price than small eggs. Fat animals will sell for a higher price than bony animals.

Where there are marketing boards for export crops there is usually an official system of grading on the basis of quality. Again, the price paid depends upon the grading of the product. Examples of crops commonly graded for sale are tea, coffee, cocoa, tobacco, cotton, rubber, groundnuts and palm oil.

Some of the factors affecting quality and grading are evenness of sample, size or shape, condition (which may depend upon maturity or decay), freedom from foreign matter, flavour of food crops and freedom from disease and insect damage.

To produce high-quality products and get a higher price it is necessary to make special efforts, and this may involve extra costs. However, many farmers could increase their profits by improving the quality of their products.

Cost of production

Costs of production include all purchases of seeds, fertilizers, sprays, animal foods, veterinary medicines and fuels and oil for machinery used up in the production of crops and the husbandry of livestock. Purchases of new machines or buildings should not be included in a list of costs because they are not used up in the course of production. However, the loss in value (depreciation) of machines or buildings (see p. 38) should be included in estimating total costs of production.

Sometimes the resources of land, labour and capital must be paid for. Wages of hired labour, rent for land and interest on borrowed money are also costs of production.

Variable and fixed costs

For purposes of farm planning it is useful to think of two kinds of costs: variable cost and fixed costs. Variable costs apply specificially to a particular enterprise. They vary according to the size of the enterprise and can be avoided if the enterprise is discontinued. If a farmer grows 2 hectares of maize instead of 1, he will need twice as much seed and fertilizer for that crop. If he does not grow any maize then no seed or fertilizer is required. Another farmer who decides to keep ten pigs instead of five will need twice as much pig feed. If he stops keeping pigs then no pig feed is required.

Fixed costs apply to the farm as a whole. They do not vary even when quite large changes are made in the number, kind and size of enterprises. It is a waste of time and may be misleading to divide up fixed costs and allocate a share to each enterprise. Fixed costs are also known as overheads.

As an example of a fixed cost we can take the cost of owning a tractor. The original purchase price may be spread over several years as a depreciation charge. The tractor may also be licensed and insured for use on public roads. Regular servicing may be carried out. These annual costs of depreciation, servicing, licence and insurance hardly vary at all, however much the tractor is used. In other words they are fixed costs. The tractor may be used on any part of the farm. It may be used to plant and weed maize or any other crop. It may be used to transport food to pigs or other livestock. Therefore the cost of owning the tractor cannot be applied to any single enterprise in particular.

The cost of owning the tractor will not vary directly with the number, kind and size of enterprises. If the farmer grows an extra hectare of maize or keeps five more pigs he will not increase the tractor costs. If he stops growing maize or keeping pigs he will not necessarily be able to avoid these costs. Of course, the costs of *using* the tractor, fuel costs in particular, are variable but the costs of *owning* the tractor are fixed.

Similarly, most machines and equipment on the farm, even hoes and cutlasses (machetes) are not used on one single enterprise and should be considered as fixed costs. A palm oil press or a coffee pulper is used on one single enterprise, but the cost of owning it does not vary directly with the amount it is used. It is more convenient to treat costs of such specialized equipment as fixed.

The wages of full-time regular staff are also 'fixed' once the staff has been hired. Even though a man may be hired to look after a single enterprise, such as poultry, his wages will not vary directly with the number of birds kept or the number of eggs laid.

If rent is paid for land this is also a fixed cost because this must be paid whatever small changes are made in the organization of the farm. Depreciation and repairs of buildings are considered as fixed costs for similar reasons.

For purposes of farm planning it is usual to separate costs in the following way.

VARIABLE COSTS	FIXED COSTS
Seeds	Depreciation of equipment and buildings
Fertilizers	Maintenance and repairs of equipment and buildings
Sprays	Wages of regular labour
Depreciation of tree crops	Rent
Fuel for machines	Interest on borrowed capital
Depreciation of livestock	
Livestock feeds	
Veterinary costs	

The peasant farmer may have hardly any fixed costs. He has hardly any equipment or buildings so these costs are very low. He does not pay wages to his family which makes up his regular labour force. He pays no rent and he does not borrow much capital. Practically all the peasant farmer's costs are variable.

Gross margins

The difference between the value of total production and the variable cost of each enterprise is known as the gross margin (Frontispiece—top).

The gross margin of an enterprise is a measure of what that enterprise is adding to farm profits. Suppose a farmer has sufficient resources to start growing maize as a new enterprise. The fixed costs of the farm will not be altered if the farmer grows the maize. Only the variable costs will be increased together with the total value of production from the farm. If the extra variable cost is likely to be less than the value of the extra production, the farmer will increase his profits by growing the maize. In fact, his profits will be increased by the value of the gross margin for this maize.

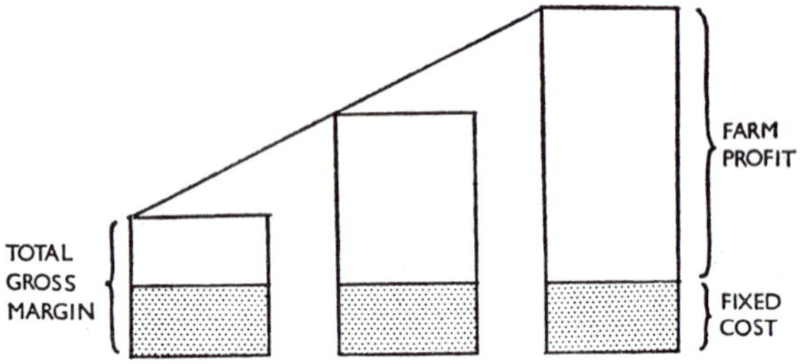

Fig. 3. The higher the total gross margin the smaller is the proportion needed to pay fixed costs, leaving more for farm profit. It usually pays, therefore, to produce as much as possible.

When a farmer has a choice between two or more possible uses of spare resources, he should choose the plan which will give the biggest gross margin. For example, if the farmer could grow either maize or cassava (manioc) on a plot of land, and he estimates that maize would give the highest gross margin, then he should grow maize.

The gross margins for the individual enterprises on a farm may be added together to give the total gross margin for the whole farm. The total gross margin is the sum out of which fixed costs must be paid. What finally remains, then, is the farm profit (Frontispiece—bottom).

Since fixed costs do not alter very much with changes in production, it is almost always true to say that, if you can raise the gross margin on a farm, you will automatically increase the farm profit. If this can be done without affecting fixed costs, the farm profit will be raised by exactly the same amount as the gross margin (Fig. 3). For this reason, it is possible to plan in terms of gross margins and leave the farm profit to look after itself. Since the peasant farmer has hardly any fixed costs, the total gross margin is almost equal to the farm profit. He can only plan in terms of gross margins.

Comparative advantage

Farming methods are not the same throughout the tropics. Farmers in different countries and even different districts specialize in different kinds of production. This is because the farmers in each district

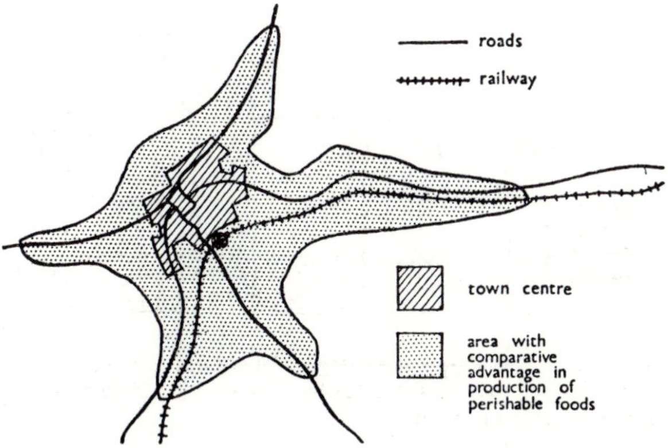

Fig. 4. Farmers near the large commercial and industrial towns have a comparative advantage in the production of fresh fruit and vegetables, milk and eggs.

produce the crops and livestock which are best suited to local conditions. In other words, they produce the enterprises for which they have the greatest comparative advantage in production. For example, the farmers in the rain forest areas have an advantage over the farmers in savanna regions in the production of valuable tree crops for export. They may also have an advantage in the production of arable food crops such as maize and cassava (manioc). Since their comparative advantage in the production of tree crops is even greater, they tend to specialize in tree crop production. Food crops are produced in savanna areas and some of these are sold to farmers in the forest areas.

An area may have an advantage in the production of a particular crop or animal because the soils, climate, topography and lack of pests and diseases make high yields possible. To get equally high yields in another area money would have to be spent on fertilizers, irrigation, insecticides and other extra expenses. Alternatively, one area may have an advantage in the production of certain goods because of its nearness

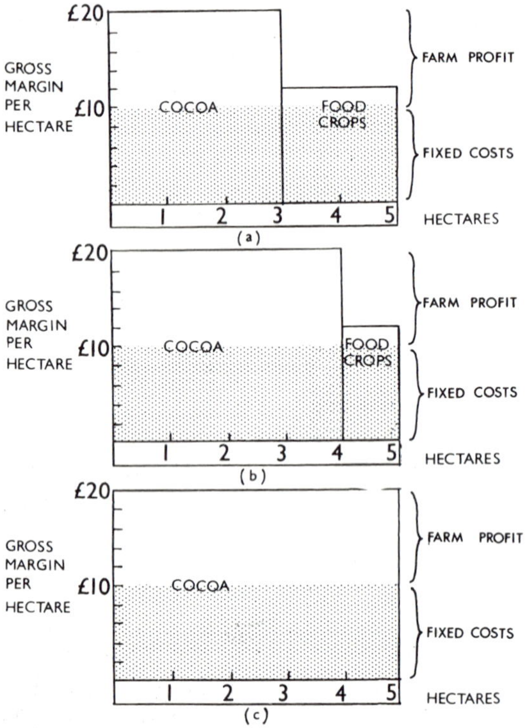

Fig. 5. If land is limited the farmer should choose the enterprise which will give the highest gross margin per hectare. In this example, if the farmer could turn 1 hectare from food crops with a margin of £12 to cocoa with a margin of £20, the farm profit would rise by £8. If he could use 2 hectares the profit would rise by £16.

(*a*) Total gross margin = £60 + £24 = £84
Farm profit = £84 less £50 = £34
(*b*) Total gross margin = £84 + £8 = £92
Farm profit = £34 + £8 = £42
(*c*) Total gross margin = £84 + £16 = £100
Farm profit = £34 + £16 = £50

to a town. There is usually a high demand and therefore a high price for fresh fruit and vegetables, eggs, milk and other perishable goods. These goods spoil quickly so they cannot be carried far except at high cost (Fig. 4). Again, there may be an advantage in producing a certain product because there is a special demand for it. If there is a rubber factory in the town, there will be a demand for rubber, and farmers round about will have an advantage in growing and tapping rubber trees. One area may have an advantage in the production of certain goods because of the supply of resources. Where there is plenty of land available farmers have an advantage in cattle production because there is plenty of grazing for the animals. Where there is plenty of labour, farmers have an advantage in the production of crops and animals which require much labour. It may be that a good supply of machinery or fertilizers or animal feeds give certain farmers an advantage in growing a particular crop or keeping a certain kind of livestock.

As far as the business farmer is concerned, comparative advantage may be measured in terms of gross margins. The enterprise for which he has the greatest comparative advantage is the one which will give the highest gross margin from his limited resources. To earn as much profit as possible he should produce as much as possible from this enterprise. If land is limited then the farmer should use all his land (if possible) to grow the crop which will give the highest gross margin per hectare (Fig. 5).

This may mean that the farmer should use all his land for sale crops and grow no food crops. He would then have to buy all the food for his family. He is correct in doing this if the gross margin per hectare from sale crops is greater than the gross margin from food crops. It is useful to remember the rule that if you can buy something more cheaply than you can produce it then it is not worth producing.

Supplementary enterprises

The amount a farmer can produce from a single enterprise is limited by his supply of land, labour and capital. When the supply of one of these resources is in full use the farmer cannot produce any more from that particular enterprise. However, he may be able to introduce a supplementary enterprise without needing any more resources. For example, a farmer may have used all his land to grow a certain crop. He may be able to grow an inter-crop between the rows of his main crop or to keep poultry without needing any extra land. Alternatively, if he is short of labour, he may be able to introduce an enterprise which uses labour during the months when he is not busy on his main crop. He is therefore able to look after the second enterprise without employing any extra labourers.

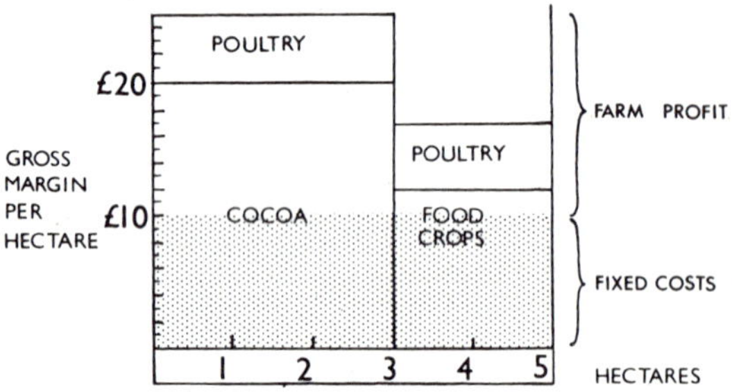

Fig. 6. Poultry as a supplementary enterprise does not compete for land. It therefore adds to the gross margin and profit per hectare of the farm. In this example the poultry adds £25 to total gross margins. Since the area is 5 hectares, poultry adds £5 per hectare to the gross margin and profit.

On many farms a supplementary enterprise could be introduced without increasing fixed costs much. Farmers should constantly look for supplementary crops or livestock which will increase the total farm gross margin without increasing fixed costs because this will lead to increased profits (Fig. 6).

Choice of methods

Having decided what to produce, the farmer should aim at finding the most profitable method of production. For example, if he has chosen to grow maize he may want to know whether to use only hand labour or whether to hire machinery. Another farmer, having decided to keep pigs, may want to know whether to feed them entirely on bought feeds or whether to feed them with home-grown crops.

Obviously the cheapest method should be used. If yields are likely to be affected, then both costs and returns must be considered. The method which gives the highest gross margin should be chosen.

Diminishing returns

Fig. 3 shows that because certain costs of production are fixed the more a farmer produces the more profit he is likely to earn. In fact, if the total value of production is very low the total gross margin may not be sufficient to pay the fixed costs and the farmer will make a loss. However, as a farmer produces more and more with a given supply of resources, a stage will be reached when the extra additions to the total amount produced will start to diminish. The use of fertilizer

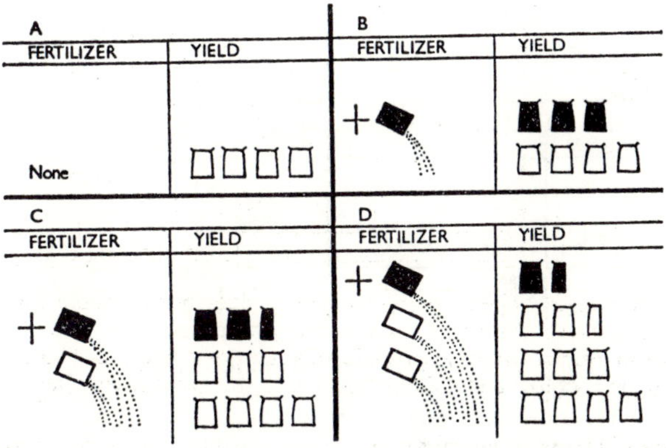

Fig. 7. Diminishing returns. (*b*) The first unit of fertilizer applied to a hectare of maize may increase the yield by three bags. (*c*) The second unit may increase the yield by only two and half bags. (*d*) The third unit may increase the yield by only one and a half bags, and so on.

on the maize crop may be taken as an example, where the land and the crop represent the fixed supply of resources. The second unit per hectare of fertilizer applied to the crop will probably add less to the total yield than the first unit did. The third unit will probably add less again to yield, and so on. The point is that beyond a certain stage, as more and more fertilizer is added, the amounts by which yield is increased get smaller and smaller (Fig. 7).

This fact applies not only to the use of fertilizers but to all aspects of business. It is known as 'the law of diminishing returns'.

On the farm as a whole, as more and more is spent on production (as total costs are increased) at some stage the amounts by which production is increased start to get smaller. While the value of the extra production is greater than the extra cost, more should be produced because this will lead to extra profit. As returns diminish, a stage may be reached beyond which the value of the extra production is no more than the extra cost. Further increases in production are

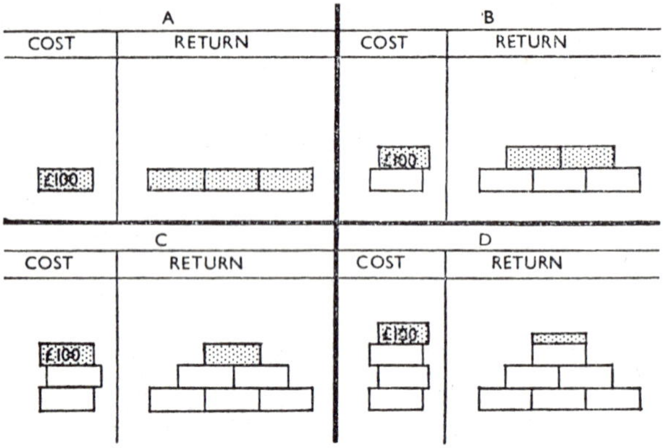

Fig. 8. Diminishing returns on the whole farm. (*a*) The first £100 spent on the farm may produce £300 return. (*b*) The second £100 may produce only £200. (*c*) The third only £100. (*d*) The fourth only £50. In fact the fourth £100 should not be spent because the total farm profit would be reduced.

not worth while (Fig. 8). Remember that production should not be expanded to the point where the value of extra production is not sufficient to cover the extra costs.

In practice this point is rarely reached and it is probably safe to say that each section of the farm should be big enough to cover the fixed costs and may be expanded as far as the limited resources will allow.

3 Farm Land

The importance of maintaining the productive powers of the soil

Land is different from other resources in that it cannot be created by man. Its qualities have been called 'the original and indestructible powers of the soil'. In the tropics, however, it is obvious that the powers of the soil *can* be destroyed. Under intense tropical rains soil will be washed away or eroded if left uncovered, even on the gentlest of slopes. In some drier areas wind is the chief agent of soil erosion. It is not so much the amount of rain falling during the year that determines the degree of soil erosion of a region, but rather the intensity with which it falls. If the surface soil is eroded uniformly it is known as sheet erosion. If deep narrow gullies are formed it is known as gully erosion. As a result of bad farming practices which encourage soil erosion, farm lands of great value have been and are being lost to the world. A farmer's profit will soon start to fall if he allows his land to become eroded. He must therefore use methods of soil conservation to prevent erosion.

The productive powers of the soil also depend upon the supply of plant nutrients. Continuous growing of one crop (monoculture) soon reduces the supply of plant nutrients in the soil. In other words, soil fertility falls unless plant nutrients are added to the soil either by fallowing or by using fertilizers and manures.

In some areas land has been exploited and large profits have been made quickly by continuous monoculture. Soil conservation and maintenance of fertility add to a farmer's costs of production. Nevertheless these costs cannot be avoided if the farmer wants to go on earning profits from his land for many years.

Shifting cultivation

Before land is settled it is usually freely available to anyone to use as he wishes. There is no need to maintain the productivity of the land because, if yields start to fall, the cultivators can move on to new land which has not been farmed. On the edges of the hot deserts of the world nomadic tribes keep livestock and travel from place to place to find grazing land for their animals. In the areas of dense tropical forest there are other tribes who practise shifting cultivation. These people clear an area of forest, usually by setting fire to it, plant their crops there for 1 or 2 years and then move on to a fresh area of forest.

Nomadic stock rearing and shifting cultivation are possible only where there is an adequate supply of land available. As the population grows and land is used for various kinds of economic development these methods become unsuitable.

The extended bush fallow system

In most parts of the tropics the farming population is settled, and what is known as the bush fallow system (or land rotation) is used.

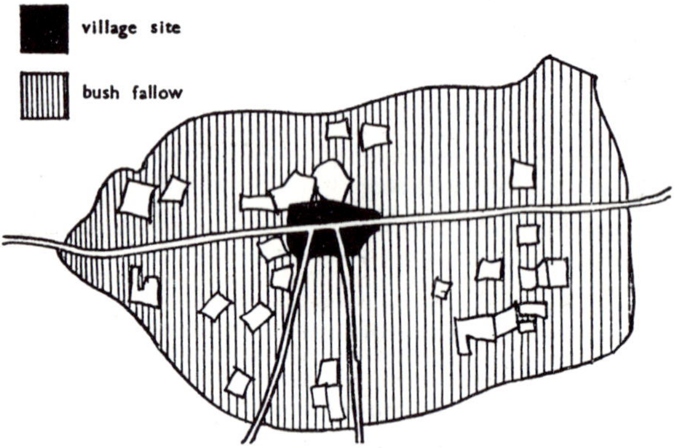

Fig. 9. The extended bush fallow system. This plan shows the village boundary clearly defined. The white areas are plots which are cropped this year. Many of these will go back to bush fallow next year and fresh areas will be cleared.

A portion of land is cleared and crops are planted on it for a few years. Then it is allowed to go back to bush while a fresh area is cleared for planting. The boundary of the land farmed by the villagers, however, is clearly defined and after several years of fallow the land which was allowed to go back to bush is used again to grow crops (Fig. 9).

The bush fallow period allows soil fertility and productivity to build up, provided it is sufficiently long. Where possible, it is usually from 7 to 10 years. As all the land is fallowed from time to time by this system, the productivity of the whole farm is maintained and soil structure restored.

The settlement of land was never planned from the beginning by any government, so the resulting sharing-out of the land was not always fair nor well organized. As villages grew the land often came under communal ownership. This meant that the land belonged to the village as a whole and not to one individual, although each family was allowed to farm a share of the village land. Communal ownership of land is still common in many parts of the tropics.

The problem of small scattered holdings

As the children grow up in a farming family the father usually allows his older sons to farm some of the land. The father's farm

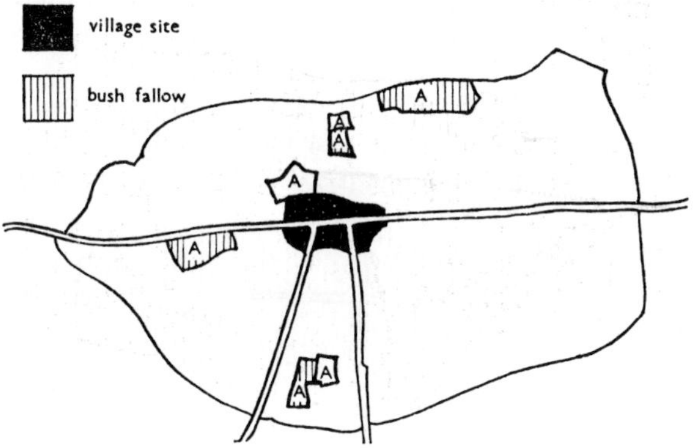

Fig. 10. Plan of the same village showing the many small scattered plots belonging to a single farmer (marked with letter A). The unshaded plots are those under current cultivation.

is thus shared amongst the family. This of course means that farms get smaller. If farmers have large families and cannot get new land this can be a serious problem. But splitting up of farms can be avoided if the children leave home. The movement of young people to the towns or to new areas of settlement may be a good thing from this point of view. As farms become smaller, land becomes a serious limitation on farming and one result is that farmers reduce the length of the bush fallow. This eventually leads to a fall in yields and farmers are worse off than before.

In many villages the land is shared out so that each farmer has some good land and some poorer land, some upland and some river valley land, some land suitable for tree crops, some land suitable for vegetables. This may be a fair system but it often means that each farmer has several small plots scattered in different parts of the village farming area (Fig. 10). Small scattered plots are not suitable when machinery is to be used and are wasteful of labour. Much time may be spent unproductively in travelling from one plot to another. Tools, seeds and harvested crops must be carried a long way back to the house.

It is also difficult for the farmer to supervise the labourers' work on his crops. It is difficult for him to see that the plants are growing well and not being damaged by animals, insects and diseases or stolen by thieves. It would be much better if each farmer had his farm in one single unit and built his house on it so that he could live there.

Improving farm layout

For agricultural improvement and mechanization it is important to have larger holdings all in one piece. Farmers can often do a great

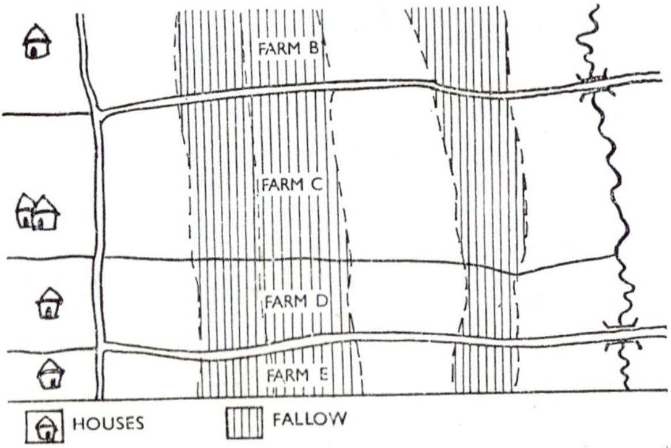

Fig. 11. Results of land consolidation. Each farmer gives up his small scattered holding and receives in exchange a farm, properly laid out, with land all in one piece and with suitable access roads.

deal to improve the situation. If they all agree to co-operate, the land can be re-allocated on a planned basis. Each farmer must give up his present scattered holdings and receive in exchange land in one piece amounting to the same area but properly laid out. This re-allocation of land is known as 'land consolidation' (Fig. 11). Communal lands may also be divided up amongst the farmers.

The measurement of land and the planning required for consolidation involves a lot of work, and technical advice may be necessary. Governments may encourage and help land consolidation schemes, as is the case in East Africa. In other parts of the tropics, such as the Western Region of Nigeria, the government has tackled the problem by resettlement. For this purpose the government actually acquires land which is laid out in well-planned holdings worked on a co-operative basis by the settlers.

These government projects can only affect a few farmers. The rest must try to improve farm layouts by their own efforts.

Acquiring more land

In some parts of the tropics land is still freely available but it is often in remote areas. People sometimes do not like to leave their homes to farm new land. In many other parts there still exists communally owned land not yet farmed. Very often this land may be obtained by a simple request to the village head or council.

Today more and more land is being bought and sold or hired. Where land is hired, the money paid for its use is called rent. Even when produce is given to the owner in return for the use of the land, it can still be considered as rent.

Rent and sale value

The rent for a piece of land is much less than the sale value because rent is paid every year whilst the sale price is paid once and for all.

The rental value or the sale value of a piece of land must depend upon how much gross margin can be produced per acre. If the land is very productive and has a comparative advantage in the production of a valuable crop, then the rent and sale value will normally be high. It is very important for a farmer to consider what he himself could earn per hectare before arranging to buy or rent land.

When land is limited

Parts of West Africa and many Asian countries are thickly populated. Farms are very small and it is impossible to acquire more land. In these conditions land is the resource which limits the amount farmers can produce. It is therefore important to earn as much as possible from the land which is available.

Land is normally measured in hectares. The aim, where land is limited, should be to earn as much gross margin as possible *per hectare*. Some tree crop farmers worry about their production and costs per tree. Arable farmers may worry about the cost of producing one bag of sorghum or one basket of yams. It is much more useful to think in terms of production and costs per hectare.

To earn high gross margins per hectare of the whole farm any of the following methods may be used.

(1) Grow more profitable crops.
(2) Practise multiple cropping, provided this does not reduce yields.
(3) Shorten the length of the bush fallow, provided this does not reduce yields.
(4) Keep livestock, where they do not compete with cash crops.
(5) Increase yields.
(6) Aim for high prices.

Grow more profitable crops

This is an application of the principle of comparative advantage. If land is the limiting resource, produce as much as possible of the crop which will give the highest gross margin per hectare. A farmer should consider whether he could introduce another crop which would give higher gross margins per hectare than the crops he is already growing. If the answer is no, he should then consider whether he could increase the area of the crop which is giving the highest gross margin per hectare.

Practise multiple cropping

Throughout the tropics inter-cropping is practised. In a yam plot other crops such as maize, cassava and vegetables may also be grown. Although tree crops are often grown alone, rice is probably the only arable crop commonly grown without a mixture of other crops.

Mixed cropping usually leads to higher total production per hectare than a single crop does, except in some semi-arid regions. Each individual crop may yield less in a mixture, but the total production is usually higher (Fig. 12). It is difficult to use machinery in mixed crops so, if a farmer wishes to mechanize his crop production, he should avoid this method of increasing total gross margin per hectare.

Fig. 12. Multiple cropping can increase total yield per hectare. Groundnuts and maize in a mixture produce less than they would as a sole crop. Even so the total yield from 1 hectare of groundnuts and maize mixed is greater than the yield from 1 hectare of groundnuts or maize alone.

Besides inter-cropping, many farmers grow more than one crop during the year. An early crop is grown and this is followed by a late crop. This kind of multiple cropping also leads to a higher total production per hectare than a single crop does, although each individual crop may yield less per year.

When multiple cropping is practised the crops are supplementary to each other in the use of land.

Shorten the length of the bush fallow

A farmer will normally have some of his land resting under bush fallow. Clearly, the less bush fallow he has the more land he has available for cropping. In this way he is able to produce more from the land that he has. On average he may produce more per hectare.

However, shortening the bush fallow as we have seen, very often leads to poor yields unless some manures or fertilizers are used. There is therefore no value in increasing the land available for cropping if yields will be less. It is safe to reduce the length of the bush fallow only provided that yields are kept up by improved methods of crop production (see Chapter 6).

Keep livestock

The value of livestock products—meat, milk and eggs—is usually greater than the cost of the food the stock eat if they are well managed. For this reason a farmer can earn more money by feeding his crops to livestock and selling their products than he can by selling the crops directly. Pigs and poultry may be fed on purchased feeds so that they require hardly any land. They add to the total gross margin from the farm and therefore to the gross margin per hectare.

The manure produced by the livestock can be used to increase soil fertility so that crop yields are raised. Other advantages of keeping livestock on a crop farm are given in Chapter 7. The only problem is that capital is needed to buy livestock, and capital is often limited.

On many farms the major export crops may earn a higher gross margin per hectare than feed crops grown for livestock. Where this is the case the export crop should be grown.

Increase yields

Apart from all these suggested changes in the number, kind and size of enterprises it is important to produce high yields. High yields of crops and animals may often be achieved at very little extra cost. No extra land is required, so high yields can be expected to give high gross margins per hectare.

The techniques of good husbandry discussed in Chapters 6 and 7 are concerned with producing high yields and high gross margins from crops and livestock.

Aim for high prices

Returns per hectare may be increased, by selling produce at higher prices. These may be obtained by the methods (described in Chapter 2) of selling at a special season or producing better quality crops and livestock.

4 Farm Labour

The importance of family welfare

The resource called labour is the work done by human beings and not the persons themselves. When a farmer hires a labourer he is buying only so many hours of work and not the man himself. Labour is actually more than just another resource under the farmer's control. To get the best out of workers they must be treated as individuals and their welfare considered. Good labour relations are necessary.

On most small farms in the tropics no labour is hired. The work is done by the farmer, his wives and his children. For this reason he may not think of labour as an important cost. However, the welfare of the family should be even more important than the welfare of hired workers. In fact, the real aim of most farmers is to make a good life for his family; to keep them happy and well. The purpose of earning profits is to provide money for the family's welfare.

The farmer's family should get enough of the right kinds of foods. Root crops and cereals contain carbohydrates which provide energy for working. But other foods are necessary, especially for growing children. Proteins are essential for healthy growth and development. These are contained in milk, meat and eggs and peas and beans. Vitamins are also necessary, and these are contained in fresh fruit and green vegetables. Children in particular should get plenty of these foods. The farmer should be prepared to buy or else to produce enough of the right kinds of foods for his family. At the same time he must spend money for medical attention to keep them fit and well. Furthermore, a farmer should not work his family too hard. There is no point in earning high profits if each member of the family has to work 12 hours a day. It is important to have time for other activities. In particular, the children need time for education. The farmer himself should try to find some spare time for sitting down and working out plans for his farm. Time is also necessary for eating and sleeping. A normal working day in many parts of the world is 8 hours, although at certain times of the year farmers may work longer than this.

Family labour is limited

Where no labour is hired the amount of labour available for work is limited by the size of the family. In many areas the supply of family labour is getting less. In some places farmers are taking fewer wives and having less children. But more important than this is the tendency for wives in some countries to take up other jobs such as trading, for children to spend more time at school and for the older children to leave home.

In many parts of the tropics the area of land cultivated is decided by the amount of work the family can do. If the farmer could get more labour he could cultivate more land. In such a case the labour supply is limiting the size of the farm.

Farm work, as is well known, does not occur evenly month by month through the year. There are certain busy periods at planting time and harvest. It is at these busy periods, also called 'work peaks', that the labour supply can be a serious limitation. Work peaks occur because there is not much time in which to complete the tasks. If planting or harvesting is delayed there will almost certainly be a loss in yields. It is essential to finish these tasks quickly, but this means that extra labour is required at these particular times of the year. Fig. 13 shows the work peaks on a farm in Trinidad in July, August, February and March. At other times of the year there may not be sufficient work to keep the whole family or farm staff busy, e.g. May and June in the example given. It is probable that the farmer will either have less labour than he wants on the farm at work peaks, or more than he wants at slack times, or something of both.

One solution to this problem is to hire temporary labour only at the busy periods, but this is not always possible. Alternatively, members of the family may be able to organize their household duties and other tasks so that they are free to help on the farm at times when this help is most valuable. It may be necessary for members of the family and hired workers to work extra long hours at the work peaks. Whatever method the farmer may use to solve the problem, the amount of labour available at the busy periods must limit the total production from the farm. Labour efficiency may be improved by spreading the labour needs more evenly through the year. The labour supply at busy

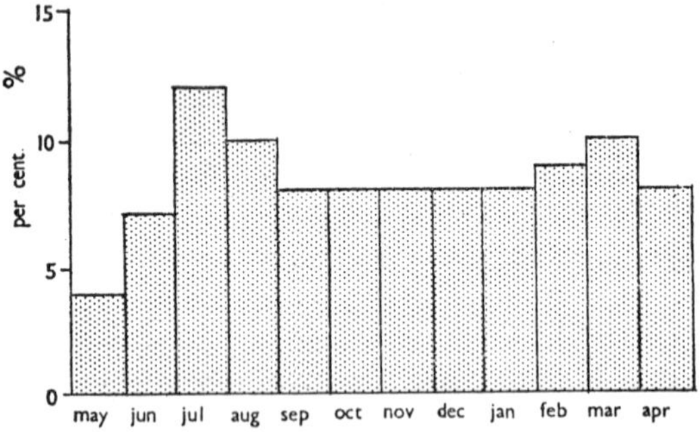

Fig. 13. The seasonal variation in man-hours (by percentage of total) worked from month to month on a 1·2 hectare farm in Trinidad.

periods may be increased by using communal labour, by share cropping or by hired labour.

Communal labour

In many parts of the world it is common for the people of a village to work together when there is a particular job that needs doing quickly. The labour of the villagers working together is known as communal labour and it may be used to build roads, schools or churches for the community. This is a good way for villagers to help themselves in improving their living conditions.

Sometimes an individual farmer is helped by communal labour to build a house or even to clear land or do some other farm work. The farmer who is helped may give food and drink to the others and this, of course, is a kind of payment.

Sharecropping

Under this system a farmer makes another man responsible for a particular crop, which the latter cultivates and harvests. He is not paid a wage but is allowed to keep a share of the crop yield. The share is a previously agreed proportion of the total crop, often a half. The other part goes to the owner.

The sharecropper, unlike an ordinary hired labourer, has a personal interest in getting high yields. Sharecropping is common in growing the rubber crop of Mid-Western Nigeria. Ibo men from the Eastern Region tap the rubber and the yield is shared equally between the farmers and the tappers.

A disadvantage of this system is that the sharecroppers may cause permanent damage in trying to get as high yields as possible. Many rubber trees in Mid-Western Nigeria have been spoiled by tapping too frequently. Much of the cotton crop in the south of the United States of America was also produced by sharecroppers. In some States soil became infertile and even eroded because they exploited the land by growing too much cotton. Similarly, the 'batai' system in India is a form of sharecropping, and this too, has led to some land being exploited.

Hired labour

Labour may be hired on a permanent basis as full-time regular labour, or on a temporary basis as part-time or casual labour. Full-time regular workers may be more skilled and better at their work than temporary hired labourers. They are also available throughout the year. However, it is usually more convenient and cheaper in the long run to hire workers when they are needed at busy times of the year. Sometimes workers are hired because they have a special skill. For example, in the Busoga area of Uganda where the local farmers have little interest in cattle, professional herdsmen are employed to look after the communal herds.

Temporary labourers may be paid a daily wage or they may be paid according to the amount of work done, when it is known as piece-work. Whereas the daily paid worker is required to work for a definite number of hours each day, the piece-worker can go home when his fixed amount of work is finished. The piece-work system seems the more just method of payment since the worker is paid directly according to the amount of work he has done. However, it may be more difficult to organize since it means that the farmer must decide new rates of pay for each task which is done.

Labour costs

Some governments make laws stating the minimum wage rates which must be paid to hired labourers. Where this is not the case wages must be decided by the farmer bargaining with the workers or their trade unions. Similarly, the proportion of the crop received by the sharecropper must be decided by bargaining.

The labour costs of a farm are normally represented by the wages of hired labour. However, in some cases members of the family working on the farm could be earning money in some other occupation. Wives could be trading, for example. By working on the farm they are going without the money they could have earned in another occupation. The money that could have been earned may be thought of as a cost, and called the 'opportunity cost' of family labour. Often the amount of work on a farm is too much for the farmer and his family so he must use extra labour. However, some farmers do not want to work because they think it is undignified or because they are lazy. These farmers should remember that the costs of hired labour or the value of the part of the crop taken by the sharecropper could have been extra money in their own pockets. They make themselves poorer by hiring labour or by sharecropping.

When labour is limited

Where land is plentiful the size of farms is limited by the labour supply. The shortage of labour is often accompanied by a shortage of capital. In these circumstances the farmer should try to earn as high a gross margin as possible with the available labour.

The unit of labour is not 'one man' but rather the work of one man for a specified period of time: for example, 1 man-hour. Two man-hours may be the work of two men for 1 hour or one man for 2 hours. The total number of man-hours is obtained by multiplying the number of men by the number of hours worked. Generally speaking, a woman or a child cannot do as much work in 1 hour as a man can. The number of hours worked by women and children may be multiplied by $\frac{2}{3}$ and $\frac{1}{3}$ respectively to convert them into man-hours (Fig. 14).

The average gross margin per man-hour from the whole farm can be increased either by producing more with the present labour supply or by saving labour. A saving in labour may mean simply that the farm staff work fewer hours. This kind of saving will not increase profits. To make a saving in costs the farmer must actually employ less labour. It may be easy to stop hiring temporary workers but it is often difficult to manage with one less regular worker. If no labour is hired a farmer may find it impossible to find outside work for one of his family. In this case the aim must to be to produce more.

Fig. 14. Two man-hours may be the work of one man for 2 hours, one woman for 3 hours ($\frac{2}{3} \times 3 = 2$) or one child for 6 hours ($\frac{1}{3} \times 6 = 2$). Alternatively two man-hours may be the work of two men for 1 hour, three women for 1 hour or six children for one hour.

The following are ways in which the gross margin per man-hour can be increased.

(1) Produce more per hectare.
(2) Choose the right enterprises to produce.
(3) Improve farm layout.
(4) Use improved tools and working methods.
(5) Keep good labour relations.
(6) Pay wages in relation to the amount and quality of the work done.

Produce more per hectare

The amount of regular labour employed per hectare of the farm represents a fixed cost. Generally high gross margins per hectare can be produced without much more labour than is needed for low gross margins per hectare. Besides improving efficiency of land use, increased yields and mixed cropping improve efficiency of labour use. Very little extra labour is involved in producing high yields. The amount of work done in cultivating, planting, weeding and harvesting a crop is usually not affected by the yield. A hectare of mixed crops can be grown without much more labour than a hectare of a single crop requires.

Choose the right enterprises to produce

Choose those crops and animals that earn a high gross margin per hour of work. These are not necessarily the same enterprises that earn most per hectare because some high-yielding crops and stock need a lot of labour.

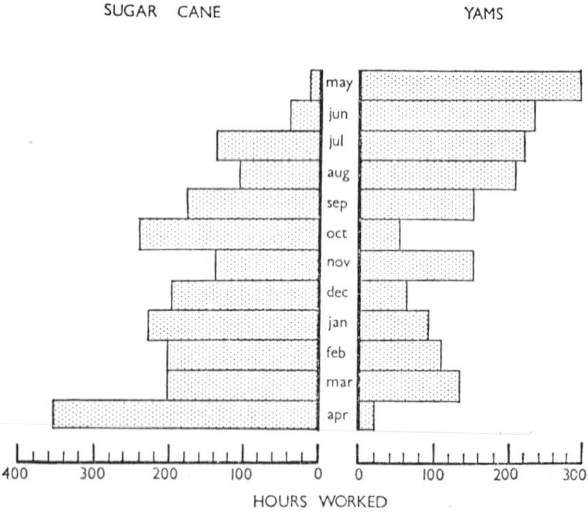

Fig. 15. The monthly distribution of labour on crops (excluding grass).

Another point to remember in deciding what to produce is that different enterprises need labour, as we have seen, at different times of the year. Fig. 15 shows how on the farm in Trinidad the work peak for sugar cane occurred when there was not much to do on the yam crop and the work peak for yams occurred when there was not much to do on the cane crop. The different work peaks of these two crops therefore balanced each other. As a result the total labour need of the farm was fairly even throughout the year.

The farm was 1·2 hectare and the crops originally grown were yams, soya beans, maize, pigeon peas, sweet potatoes, and sugar cane. This cropping system led to big variations in the labour needs from month to month, as was shown in Fig. 13. The farmer and his wife, who between them did all the work, were clearly very busy in the month of July but had little to do in May.

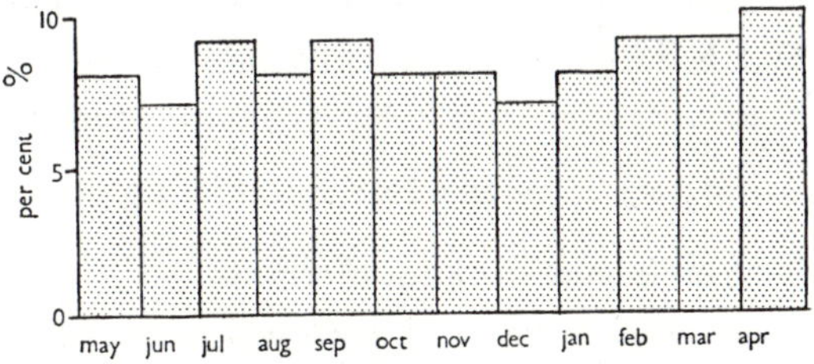

Fig. 16. The seasonal variation in man-hours (by percentage of total) worked after replanning the 1·2 hectare farm in Trinidad.

The farm was replanned so that the only arable crops grown were yams and sugar cane. As a result, the total labour requirements became much more evenly spread over the year (Fig. 16). On many farms it should be possible to find combinations of enterprises which supplement each other in the use of labour in this way, so that more can be produced with a given labour force and the productivity of labour thus increased.

Improve farm layout

The importance of good farm layout has already been mentioned in discussing farm land. A large single plot is much better than many scattered plots because less time is wasted in travelling to and from work. For similar reasons a farm which is compact, with the buildings near the centre, is better than one which is long and narrow with buildings at one end (Fig. 17).

Effort spent in improving farm layout is usually well worth while since it leads to big savings in labour, quite apart from other advantages.

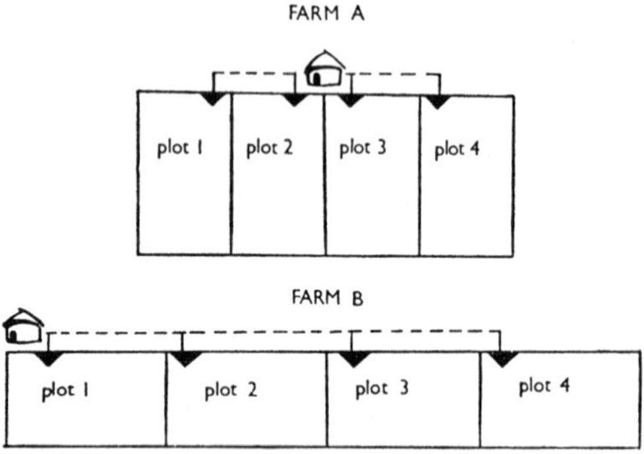

Fig. 17. The importance of farm layout. Although the two farms shown are the same size, the distances travelled (shown by the dotted lines) to reach the plots on farm A are far less than the distances travelled on farm B.

Use improved tools and working methods

Labourers can work better with well-kept tools. Farmers should always make sure that hoes and cutlasses are kept sharp and that the handles of tools are fitted firmly. In many cases labour efficiency could be improved simply by keeping tools and equipment in better condition.

There are many different kinds of farm tools and machines in use in the world. Some of these have been common in certain areas for a

Fig. 18. The wheelbarrow. It should have a large wheel preferably with a rubber tyre to ride over bumps. The load should be arranged over the wheel as far as possible, to take the weight off the handles.

very long time; others are fairly new inventions. Most farmers could improve the productivity of their labour by adopting better tools which have been tried out elsewhere.

As an example, one piece of equipment which can improve labour efficiency is the wheelbarrow (Fig. 18). Loads are often carried on the head or on the back so that they can be taken through the bush with ease. Where there are paths or roads with an even surface, however, a man with a wheelbarrow can transport about three times as much as a man can carry.

Fig. 19. The scythe in use. The wire cradle is attached to catch the cereals as they are cut. When cutting green crops the cradle is removed.

Another tool which might be worth trying is the scythe (Fig. 19). It can be used for cutting grass and other green crops and for harvesting cereals. Since it cuts near to the ground, the surface must be flat. It is claimed that green crops can be cut five times faster with a scythe than with a sickle or machete, and cereals can be reaped three times faster. A progressive farmer will not hesitate to try such new tools if he thinks they will help him and his labourers to speed up their work.

A considerable amount of labour is involved in processing crops after harvest, as many women know. Fortunately, mechanization of this processing can now save a lot of labour. There are machines available for shelling groundnuts, pressing palm oil fruits and cracking the kernels, threshing grain, hulling coffee and rice, and for grinding. These machines not only save labour but also give a better quality product.

The productivity of labour may also be improved by the use of animals such as cattle, horses, donkeys or camels. These animals can pull bigger implements than a man can manage alone. In tropical rain forest areas, where livestock often suffers from disease, a tractor or a motorized implement is the only alternative to hand-labour for field work. Since tractors can work so much faster than men with hand-tools they naturally save labour.

It must be noted that keeping working livestock and tractors and machinery is an additional cost. A farmer who is thinking of buying a working animal, a new machine or a piece of equipment, therefore, should make sure that the saving in labour or the extra returns from increased production will be sufficient to cover the extra costs.

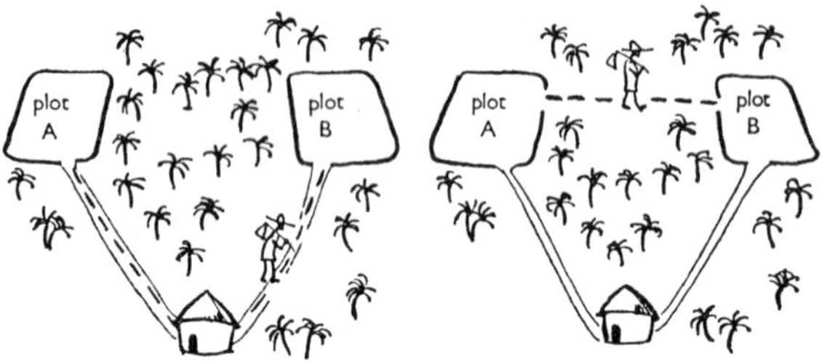

Fig. 20. How a new path can save distances travelled from one plot to another. Distance travelled from plot A to plot B is shown by the dotted line.

It is certain that the costs of owning and running a tractor are normally too great for an ordinary farmer to afford. On the other hand, it may be possible for him to share a tractor with others or hire the use of one. The costs of keeping working animals are generally less than those of owning and using a tractor.

The introduction of improved tools and machines usually means a break with tradition. New tools require new working methods and skills. For example, the use of the scythe needs considerable skill and practice. The farmer and his workers must be prepared to learn the new methods. Even when *no* new tools or machines are introduced, new working methods can be found which save labour. Making a new path to cut down the distance walked from one part of the farm to another is one way of saving labour (Fig. 20).

In many industries, including agriculture in some parts of the world, the study of working methods has become very important. It is known as work study or time and motion study. Normally only farmers who run big farms can afford to employ work study specialists, but even on a small farm it is worth while to look for ways of doing jobs more quickly and easily and of reducing wasted time.

Keep good labour relations

The importance of good labour relations has already been mentioned. It is no use treating labourers as slaves. They must be chosen carefully and treated justly. Good supervision is important particularly with hired labour. Regular or daily-paid labourers may not work hard unless they are watched from time to time. Piece-workers often do not work carefully unless they are watched. Naturally, the farmer cannot always watch his workers and, at times, must trust them to work alone. But good supervision usually leads to better results.

Pay wages in relation to the amount of work done

Piece-workers are paid in relation to the amount of work done. However, it is suitable only where the amount of work done depends upon the physical effort of the workers. It is not suitable where the worker does many different tasks during the course of the day, because new wage rates must be worked out for each task.

Some farmers pay their workers extra money, known as a bonus, according to the amount produced. For example, a farmer may pay the labourers on his rubber plantation a bonus for every kilogramme produced. Cowmen are sometimes paid a bonus per litre of milk produced.

There is a danger in paying bonuses because the worker may be concerned with producing as much as possible without paying attention to the costs, as in sharecropping. A cowman may give the cows too much food in order to get high yields and a high bonus. Rubber workers may tap the trees too often.

Another way of encouraging workers to give of their very best is to offer them a share of the yearly farm profit. This may be difficult to arrange because the total farm profit often cannot be measured accurately. There is also the danger that the workers will want to help manage the farm if they expect a share of the profits. A farmer must therefore make a careful decision about how to encourage his workers with money incentives.

5 Farm Capital

What is capital?

Most people have an idea of what capital is, but would find it difficult to put it into words. Capital is made up of things which have been produced but are not yet used up. Capital includes not only buildings and machinery, roads and footpaths, but also drainage ditches, terraces, growing tree crops, animals, stocks of seed, and so on. It should be clear that with extra capital almost any farmer could earn more profit for himself. In fact, we can say that it is impossible to start farming or to improve a farm without some capital. If a farmer wants to grow a few more acres of cocoa or keep a few poultry he will need capital—or a supply of money to buy it. He needs money to buy a piece of equipment—for instance, an oil press or a coffee pulper.

Capital depreciation

Capital has been described as things which have been produced but not used up. Of course everything gets used up or worn out eventually, but capital is normally being replaced all the time. When a hoe is useless a new one is bought; when a laying hen dies it is replaced. The replacement of worn-out capital must cost money. If the capital lasts more than 1 year then it would be wrong to say that all the replacement cost occurred in 1 year. Suppose a farmer buys battery cages for poultry for £100. If these cages last 10 years it is wrong to say the costs of production in the year he bought the cages should include all the £100. It is better to assume that the cages cost £100 divided by 10 each year. That is, the yearly cost of having these cages is £10. This yearly cost is called depreciation (Table 1).

It may be better if the farmer does actually save the depreciation cost each year so that when he has to replace the capital he will have saved enough money. We have estimated depreciation very simply by dividing the total cost of the cages by the length of their useful life in years. Another way of calculating depreciation would be to assume

TABLE I

How Depreciation Spreads the Cost of Expensive Equipment

Year Number	*Actual Cost*	*Method 1* Depreciation Allowance $\frac{1}{10}$ of original value	*Method 1* Value at end of year	*Method 2* Depreciation Allowance $\frac{1}{5}$ of value at end of previous year	*Method 2* Value at end of year
1	£100	£10	£90	£20	£80
2	—	£10	£80	£16	£64
3	—	£10	£70	£13 approx.	£51
4	—	£10	£60	£10 ,,	£41
5	—	£10	£50	£8 ,,	£33
6	—	£10	£40	£7 ,,	£26
7	—	£10	£30	£5 ,,	£21
8	—	£10	£20	£4 ,,	£17
9	—	£10	£10	£3·5 ,,	£13·5
10	£100	£10	£100+£0	£2·5 ,,	£100+£11
11	—	£10	£90	£22 ,,	£89
12	—	£10	£80	£18 ,,	£71

and so on

Fig. 21. Good maintenance reduces depreciation.

Tractor A is poorly maintained and lasts 5 years. Annual depreciation $= \frac{£500}{5} = £100$.

Tractor B has good maintenance and lasts 10 years. Annual depreciation $= \frac{£500}{10} = £50$.

that the value of the cages falls by a certain proportion of their value at the beginning of the year. If the proportion chosen is $\frac{1}{5}$, then the cages would depreciate by $\frac{1}{5} \times$ £100 = £20 in the first year. The value at the end of the first year, or the beginning of the second, would be £100 minus £20 = £80. The depreciation in the second year is then $\frac{1}{5} \times$ £80 = £16. The value therefore falls to £80 minus £16 = £64 at the end of the second year, and so on. This method of calculation is quite useful in practice (see *Method* 2 in Table 1).

Sometimes capital goods can be sold second-hand, in which case it is easy to measure depreciation. If a farmer buys a spraying machine for £50 and sells it after 1 year for £40, the depreciation is £10.

It is possible to make capital last longer by careful maintenance This will make the depreciation less (Fig. 21).

Ways of getting capital

Many farmers seem to think that the only way they can get capita is by borrowing. They forget that they could save the necessary money or even make capital themselves. It should be remembered that capita which is borrowed must have been saved by someone else. Saving means going without some of the pleasures of life. If a man saves £1 each week he has £1 less to spend on his pleasures or comforts. It is only reasonable to expect to get something back in return for lending the capital.

Interest on loans

Usually the money-lender expects to be paid some extra money in payment for making the loan. This is known as interest. A man lending £100 may expect to get £110 back in a year's time. In other words he wants £10 interest per year. This is usually stated as a percentage of the original sum loaned; in this case £100 was loaned, so the interest rate is 10%. It will be seen that interest is the price paid for the us of capital, just as rent is the price paid for the use of land.

Security for loans

The other difficulty about lending money is that the borrower ma never be able to pay it back. There is a risk involved in lendin capital. Sometimes it is possible for the borrower to promise to giv something to the lender if he cannot pay the money. A rich man wishing to borrow money could promise to give up his car to the lender if he could not repay the money. Anything which is promised to a lender if the money cannot be repaid is called security. Clearly, if there is good security for a loan there is no risk involved.

If money is borrowed to buy land or building, the land or the buildings themselves may be offered as security for the loan. This kind of security is referred to as a mortgage.

Farmers' disadvantages in getting loans

Farmers frequently have nothing suitable to offer as security for loans. This is particularly so where the land and buildings belong to the village or clan or are rented from a landlord. At the same time, people with money to lend usually like to lend it in large quantities, probably in thousands of pounds. A peasant farmer may want to borrow only about £50, but a lender prefers to make one big loan rather than many small ones. Also, a lender usually prefers to lend his money to people who are near at hand. Farms, however, are often remote and not easy to reach. For these reasons it is not always easy for farmers to borrow money.

Sources of loans

Moneylenders. In most parts of the world there are private money-lenders but the farmer is often unknown to the lender, who therefore asks very high interest rates. It is usually not worth while borrowing capital at these high rates because the value of extra production may be less than the extra cost of this interest.

Banks. Those farmers who keep their money in the bank may be able to get loans from the bank. However, for the reasons given above bankers prefer to lend money to business men in the towns and it may be difficult for farmers to borrow money when they have no security to offer.

Government organizations. Some governments set up special agricultural credit organizations to lend money to farmers. Where there is such an organization farmers are very lucky because they can usually borrow money fairly easily for any reasonable plan, at a low rate of interest.

Co-operative credit. Where there are no credit organizations the best thing farmers can do is to help themselves. This may be done by starting a co-operative society. The farmers all put their savings together and then, if one farmer requires capital, he borrows from this general fund. The farmer who borrows should pay interest which can be paid back to the original owners of the money. In view of the general shortage of capital and the difficulty of getting loans it is obviously important to use capital efficiently.

Capital limitations

Capital is always limited. It should therefore be used where it will add most profits. In planning what to do with available supplies of this resource the farmer should choose the plan which is likely to give the highest gross margin.

On some farms capital is very severely limited and the farmer cannot acquire more. In such a case annual arable crops are the most suitable enterprises since they produce quick returns and require very little capital. Tree crops and livestock often take several years to reach productive age. During this period labour and money must be spent on them so that they will develop well. But since nothing is being produced for sale, there is no money coming in to pay the costs. The farmer who wants to produce tree crops or livestock such as pigs or dairy cows needs capital to live on while the trees or stock are growing. The purchase of livestock and the necessary buildings and equipment also requires capital savings. The young man who is just starting farming is often very short of capital. He may have to grow arable crops for some years before he can afford to start growing tree crops or keeping livestock.

Capital is made up of many different items. Tree crops, livestock, buildings and machinery are four major items of capital which are discussed separately below.

Capital in tree crops

The capital cost of, say, a cocoa tree is roughly equal to the cost of growing that tree up to the stage when it reaches full production. Eventually the tree will die so it will have to be replaced by a new tree. Trees, in fact, depreciate like other kinds of capital. It is important to take good care of the trees and control diseases and pests. If this is not done, trees may die quickly and depreciation is therefore high.

The farmer who cannot afford to grow tree crops himself may spend part of his time working as a sharecropper on tree crops belonging to someone else. In this way he is using the other person's capital.

In order to get a high return on capital or a high gross margin from capital in the form of tree crops it is essential to get high yields.

Capital in livestock

The capital cost of an animal is roughly equal to the cost of keeping that animal up to the stage when it reaches full production. If it is sold or slaughtered immediately for meat there is no depreciation cost. On the other hand, animals kept for breeding, for milk production or for work depreciate over their productive lives. Poultry kept for egg production also depreciate. To keep the rate of depreciation low it is important to take care that the livestock are fit and healthy.

Some farmers may not be able to afford to buy or raise livestock. In some areas breeding livestock can be hired under a system of livestock leasing. The farmer looks after an animal for someone else and returns half the offspring to the owner as a kind of hire charge.

In order to earn a high gross margin from capital in the form of livestock, high yields are essential. A high yield is the same as a high growth rate in the case of animals kept only for slaughter as meat.

Capital in buildings

The cost of buildings can often be kept low if the farmer does most of the building work himself. Costs of other improvements such as drainage, fencing, terracing or irrigation may be kept low in the same way. This is not possible if the farmer has already too much work. Improvements and repairs and maintenance can often be fitted in when the labour staff are not busy between work peaks.

Careful repair and maintenance of buildings will make them last longer and will therefore reduce the depreciation cost. It is often possible to hire buildings, to avoid the capital cost of purchase.

Having gone to the expense of buying or hiring a building the farmer should try to put it to as much productive use as possible. For example, if a farmer has put up a piggery he should try to arrange that it is full of growing pigs for as much of the year as possible. If a farmer has a store he should try to fill it for as much of the year as possible with products that really need storing. An empty building is a waste of money. On the other hand there is no point in filling a building unless there is some gain. Livestock should be kept in a building only if they produce more. Stores and equipment should be kept in a building only if they will depreciate less. Crops should be stored only while they are gaining in value.

There may be some advantage in the co-operative use of buildings, particularly store-houses. It is cheaper to construct one large store rather than several small ones. At the same time it is probably easier to keep the store in use if several farmers are bringing their products to be stored there.

Capital in machinery and equipment

Most machines, equipment or tools are purchased because the farmer cannot make them himself. However, it is worth spending time and effort on careful repair and maintenance because this will make the equipment last longer. Machines should be kept oiled and greased and all equipment should be kept clean and in good repair.

In some districts farmers can hire machines and equipment from the government or private contractors. There is a big advantage in hiring because it saves the farmer the cost of buying the machine. However, even when such services are available in a certain district it is not always possible for the farmer to get the use of a particular machine just when he wants it.

Machines not only save labour, they often lead to increased production. In particular, field machinery makes it possible to complete cultivation, planting and harvesting more quickly and efficiently. This often leads to increased yields.

A machine should be put to as much productive use as possible. If it is used for only a few hours each year the cost of each hour of use will be very high. An individual farmer may not be able to use a particular machine for more than a few hours on productive work. Several farmers sharing the machine could probably make far better use of it. In many parts of the world the co-operative use of machinery has been very successful. The farmers use their combined savings to buy machinery which they then take turns in using (Fig. 22). Co-operatives are discussed more fully in Appendix I.

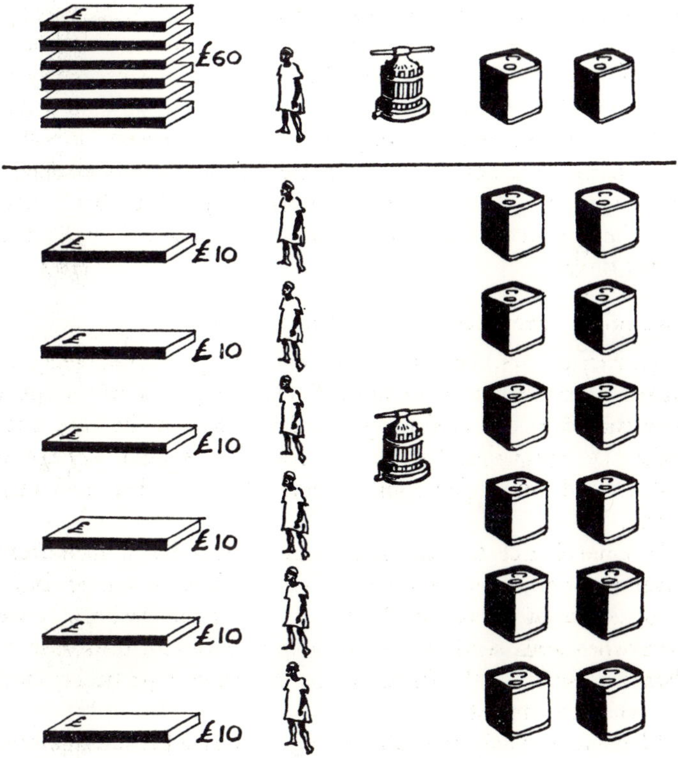

Fig. 22. The cost of a hand-operated oil press to one farmer is £60. He may use the press to extract only two tins of oil. If six farmers agree to co-operate and share this machine, the cost to each farmer is only £10 and the press is used to extract twelve tins of oil.

6 Management of Soils and Crops

The aims of soil and crop management

Once the farmer has decided which crops to grow he should aim to earn as much gross margin as possible from these crops. The method of computing the gross margin per hectare of crops is shown in Fig. 23.

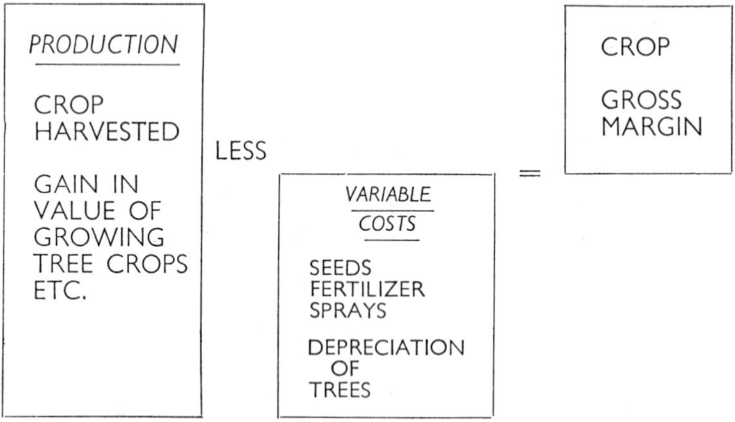

Fig. 23. Calculation of gross margin from crops.

Gross margin per hectare is increased either by increasing the crop yield and value or by reducing the variable cost. An increase in the variable cost may be a good thing if it is more than covered by the increase in value of the crop. The successful farmer puts great stress on getting a high yield per hectare. The less successful puts too much emphasis on cost reduction per hectare, without regard to the effect on the value of the crop yield. This chapter is more concerned with ways of obtaining high yields than with ways of reducing costs.

The right conditions for high crop yields

The necessary conditions for healthy crop growth are illustrated in

Fig. 24. Soil structure depends upon the presence of organic matter (humus) which is dead plant or animal material. Organic matter is also an important source of plant nutrients as it breaks down in the soil. Crop residues add some organic matter to the soil but when most of the plant is harvested and removed from the land the additions are not sufficient to maintain the amount in the soil.

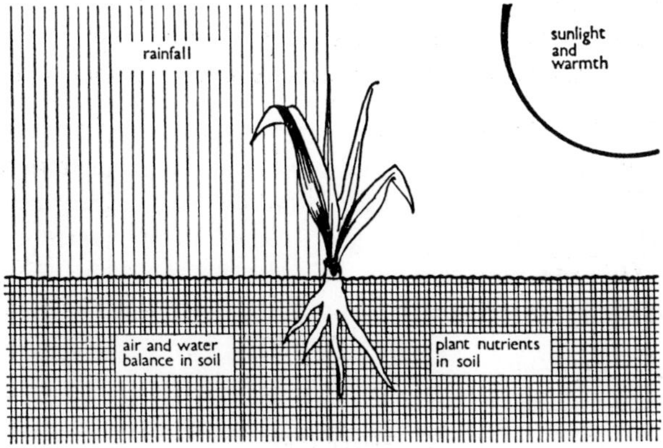

Fig. 24. The necessary conditions for healthy crop growth.

Crop plants also need sufficient space and time to grow well. Insect pests, diseases and weeds must be controlled. A damaged or diseased plant cannot grow properly and yield must suffer. Since farm crops cannot grow without soil, conservation is a first essential in management for high yields.

Soil conservation

Good soil structure. Erosion is caused by rainwater running over the surface of sloping land and carrying soil with it. This is known as 'run-off'. Conservation is aimed at stopping run-off and the washing away of soil. Where the soil structure is good, it is sufficiently firm to prevent rapid erosion, yet sufficiently porous to let water sink in rather than running off over the surface. The addition of organic matter to improve structure also helps to prevent erosion.

Keep the soil covered. The soil should be protected from beating rain, by shade trees, cover crops or mulches and should be left bare for as short a time as possible. Grassland, and crops like millets which are closely planted, protect the soil more than others which are widely spaced. Forest trees are best of all in preventing erosion.

Under widely-spaced crops such as banana, rubber or coffee the soil should be protected by allowing grass, centrosema or other creeping plants to grow. Alternatively dead plant material may be spread on the surface as a mulch to prevent erosion. A mulch also prevents the drying out of the soil during the dry period.

Land use. Run-off and erosion are worse on steep slopes than on land which is more level, so land use should be planned accordingly. Steep hill sides should be kept for forest reserves, slopes which are less steep should be used for grazing or planted to permanent crops and mulched but not clean-weeded. Annual crops should be planted only on gently sloping lowland.

Always work across the slope, never up and down. If the land is ploughed or cultivated and the crops are planted the same way as the

Fig. 25. Tied ridges. The cross bars or 'ties' of soil prevent the run-off of water and the gully erosion of the soil.

ground slopes the rain water will run down the rows and start gully erosion. If the land is worked across the slope the furrows between the rows will hold the water and it will sink into the soil or flow away very gently.

This rule is even more important if the crop is planted on ridges. These should always be constructed across the slope. If the ridges are joined by cross bars of soil (Fig. 25) they are called 'tied ridges'.

Fig. 26. Some methods of soil conservation. Note the hilltops planted with forest trees, the lower slopes with permanent crops and the arable crops on the gently sloping lowland. Note also that all the crops are planted across the slopes and that the farmer in the centre is cultivating his land across the slope.

Planting crops on tied ridges prevents both erosion and the loss of available rainfall by run-off.

Strip cropping should also be practised. The fields are divided up into strips which follow the contour line across the slope. These strips are planted with different crops, some of which protect the soil better than others. In this way the slope is broken up and the amount of run-off reduced. Some of these methods of soil conservation are illustrated in Fig. 26.

Storm drains, terraces and grass strips. Sometimes these general practices are not sufficient to prevent soil erosion, in which case physical protection is necessary. Soil can be protected by digging a storm drain across the top edge of farmed land. If the drain is large enough not to flood over the edge, it will carry the water, flowing downhill from above, gently across the slope to the nearest stream or pond. Storm drains should be shallow and wide with gently sloping sides. The sides may be protected by allowing grass to grow.

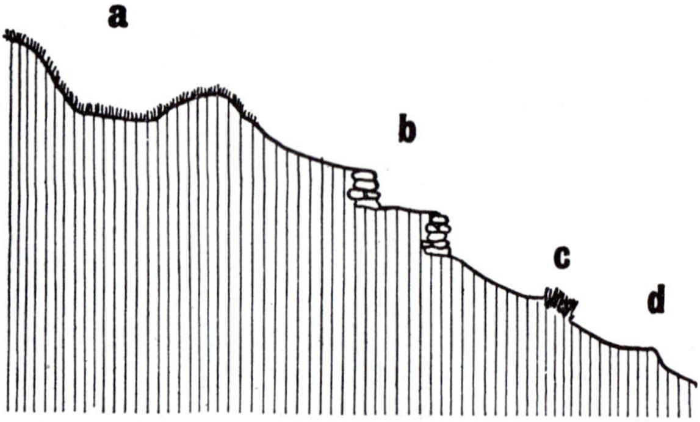

Fig. 27. Other methods of soil conservation (sectional view). (*a*) Storm drain. Note that it is across the top of the slope, it is wide and shallow with sloping sides, that the earth is thrown out on the lower side and that the slopes are planted with grass. (*b*) Terraces. Stone walls are built across the slope and the soil is allowed to build up behind the walls to form steps. (*c*) Grass strips. These hold back soil carried by run-off water. (*d*) Bunds. These are ridges of soil across the slope.

Terraces are wide steps up a hillside with the upright faces sometimes made of stones. Terracing a hillside stops the soil from being washed away, but it takes a lot of work.

On gentle slopes narrow strips of grass planted across the slope every few yards will prevent the soil being washed away. Alternatively, banks of earth called 'bunds' or contour ridges may be used for the same purpose. If properly looked after grass strips and bunds will slowly build up into terraces.

Drains, terraces, grass strips and bunds are shown in Fig. 27.

Arable crop rotation

Tree crops are grown in permanent plantations and swamp rice may be grown on the same piece of land for many years. If most other arable crops are grown continuously in the same place, without proper care, yields usually start to fall.

It is much better to grow a series of different crops, called a rotation, which may or may not include a bush fallow. Growing crops in a rotation need not add to the costs of production, but there are many advantages in using a rotation.

1. *It maintains soil structure and fertility*. Crops vary in their ability to take nutrients out of the soil. Yams will take up different nutrients from maize. Different crops also take nutrients from different levels in the soil, depending on the length of the roots. A rotation of different crops grown on a piece of land is therefore less likely to use up all the supply of a particular nutrient than the continuous growing of one crop.

Trees and larger plants which may grow during a bush fallow have deeper roots than arable crops. These plants can bring up nutrients from greater depths. When the leaves fall or the bush is cleared these nutrients are returned to the soil surface. If a bush fallow is included in the rotation the supply of nutrients in the upper part of the soil may be increased for crops which are planted afterwards. The bush fallow also improves the supply of organic matter and the soil structure. Crops such as grass, grazed by livestock, can add to the organic matter and nutrient supply of the soil.

Leguminous crops such as peas and beans actually add nitrogen to the soil. This nitrogen may be used by other crops which follow soon after peas or beans in a rotation.

2. *A rotation helps control pests and disease*. A particular insect pest or disease usually attacks one specific crop, which may be called the host crop. In the absence of a host crop the pest or disease will die out. Whereas under continuous cropping they tend to build up, in a rotation the host crop is not present every year, and this helps to control pests and diseases.

3. *It gives opportunities for weed control*. Although some weed cover may be useful in preventing erosion, if allowed to grow unchecked

weeds compete with crop plants for light, nutrients and water. Weeds must be cut down regularly to prevent this competition.

In a rotation there are likely to be better opportunities for weed control than with continuous cropping. The clearing of a bush fallow, for example, allows all weeds to be killed. Certain crops in a rotation may be more widely spaced than others. It is easier to weed a widely spaced crop.

The following advantages apply on any farm with several different enterprises. Since a rotation includes several different crop enterprises these advantages also apply to rotations.

4. *Labour requirements are spread more evenly over the year by growing a combination of different crops.* This enables men and machines to be more frequently employed on productive work. See Chapter 4.

5. *Income is spread more evenly over the year.* There are some advantages in having a steady income throughout the year, rather than just one large sum of money at a certain time of the year. If a single crop is grown and there is no storage space it may be necessary to sell all of the product at harvest time. If on the other hand a series of different crops are likely to be harvested and sold at different times, income is spread more evenly over the year.

6. *Risk is reduced.* The risk of loss from failure of a single crop is very great. It is less when a series of different crops are grown, because it is very unlikely that all crops would fail in the same year. Examples of rotations are given in Table 2.

Good crop management

Apart from the use of arable crop rotations there are other methods of good crop management which should usually be followed. These methods of management generally lead to increased yields but do not add much to farmers' costs. They therefore lead to high gross margins per hectare. Good crop management must include the return of plant nutrients and organic matter to the soil, early cultivation and planting, the correct spacing of crops and the control of weeds.

TABLE 2

Examples of Rotations

Year Number	Rotation 1 10 years Nigeria	Rotation 2 6 years Uganda	Rotation 3 3 years Japan
1	Yams i.p.w.† vegetables and pepper f.b.* cowpeas	Maize f.b. cotton	Milk vetch f.b. rice
2	Early maize f.b. cassava	Groundnuts i.p.w. beans	Barley f.b. rice
3	Cassava f.b. cowpeas	Maize f.b. cotton	Rice
4	Guinea corn f.b. cowpeas f.b. sweet potatoes	Grass ley for dairying	(repeat year 1)
5	Bush fallow	Grass ley	(repeat year 2)
6	Bush fallow	Grass ley	(repeat year 3)
7	Bush fallow	(repeat year 1)	(repeat year 1)
8	Bush fallow	(repeat year 2)	(repeat year 2)
9	Bush fallow	(repeat year 3)	(repeat year 3)
10	Bush fallow	(and so on)	(and so on)

† inter-planted with. * followed by.

SOURCES:

Rotation 2: HUTCHINSON, J., PRENTICE, A. N., FARBROTHER, H. G., LEA, J. D., and STEPHENS, A. L. (1959). The planning of a large farm for the Namulonge Cotton Research Station in Uganda, *Emp. Cotton Growing Review* 36, 81–134.

Rotation 3: SAWAMURA, T. (1963). Farm economic analysis of paddy-field cropping systems in Japan, *Rural Economic Problems*, prelim. issue.

Return of organic matter and plant nutrients to the soil. Plant or animal remains, animal manure, and ashes from the fire should not be wasted but should be spread on the land to maintain the organic matter and nutrient content of the soil. Plant materials which should be used in this way include dead leaves and stems, maize cobs, cassava and yam peelings, cocoa pods and the remains of oil palm bunches. If these plant materials are not well rotted down before they are spread, they will take nitrogen away from the soil and the crop plants at first. This does not matter if crops are not planted for some time but it is usually better to make sure the material is well rotted first.

Rotted plant material is known as 'compost'. It can be made either in a pit (in dry conditions) or in a heap (in wet conditions). A little soil or manure or old compost should be added to start the rotting process and it must be kept moist. It should be turned every 2 or 3 weeks. In 2 or 3 months the heap should be well rotted and the compost is ready to spread on the soil (Fig. 28).

Fig. 28. Compost making. All kitchen waste and unused remains from crops should be put into the compost pit. The boy is splashing water onto the compost to keep it moist. From time to time compost should be turned.

Early cultivation and planting. This almost always leads to increased yields. The land should be cultivated early enough for the crops to be planted at the very first rains. It may be harder work to cultivate the land but it should result in higher profits. Some farmers are still busy harvesting the previous crop when planting time comes and for this reason planting is delayed. It is important to get the land clear so that planting can be started early. In particular, valuable sale crops should be planted early even if this means that food crops must be harvested before they are really ready. For example, a problem arises in East Africa where maize is followed by cotton in the rotation. The maize harvest is usually well after the best planting date for the cotton. Effort made to harvest the maize and plant the cotton earlier will lead to big increases in the cotton yield.

Most tree crops should first be planted in a nursery where it is easier to look after them properly. They should be ready for planting out at the start of the main rains.

Correct spacing. If crop plants are too close together they compete with each other for light, water and nutrients. If they are too far apart land is wasted and there are dangers of erosion. For most crops, experts have discovered the best spacing to use. Farmers should find out what is the best distance to leave between the crop plants and plant them at this spacing. For some crops the seeds can be planted more closely and then some of the plants are hoed or thinned out to give the correct spacing.

Control of weeds. Whether crops are grown in a rotation or not, it is always important to control weeds. Although some weed cover may help to prevent erosion the weeds must not be allowed to compete with the crop plants. Apart from competition with crop plants weeds may also be alternative hosts for insect pests and diseases. In this way weeds can carry these pests from one crop to another.

Regular weeding especially in the early stages will almost always lead to increased yields.

Increased yields at extra cost

Although the methods to be discussed, usually lead to increased yields, they also involve extra expenses or a change in farming system. Since both value of production and costs are increased these methods may not always add to profits. If the extra cost is greater than the extra production gross margins and profits are reduced. It is important to get technical advice and to estimate the likely effect on profits and the farming system before using these methods.

New varieties of crops. Sometimes an entirely new crop which is more profitable than the old ones can be introduced on a farm. This of course requires new knowledge and the use of new production methods. However, new varieties of many traditional crops are available. The government is often responsible for the sale and distribution of such new varieties. They may yield better than the old varieties or may have resistance to insect attack or diseases. The quality of the produce from new varieties is often better. Usually the seeds or young plants cost extra but this should be repaid by the higher yields.

Very often improved varieties of crops require improved management if their use is to be really worth while. For example, improve varieties of rubber must be tapped more carefully than the old trees.

Purchased fertilizers. These are chemical salts which contain th important plant nutrients. (Fertilizers applied to crops almost alway increase the yield but they are expensive to buy.) The extra cost ma be greater than the value of the extra yield in some cases. (See sectio on *Diminishing returns* in Chapter 2.)

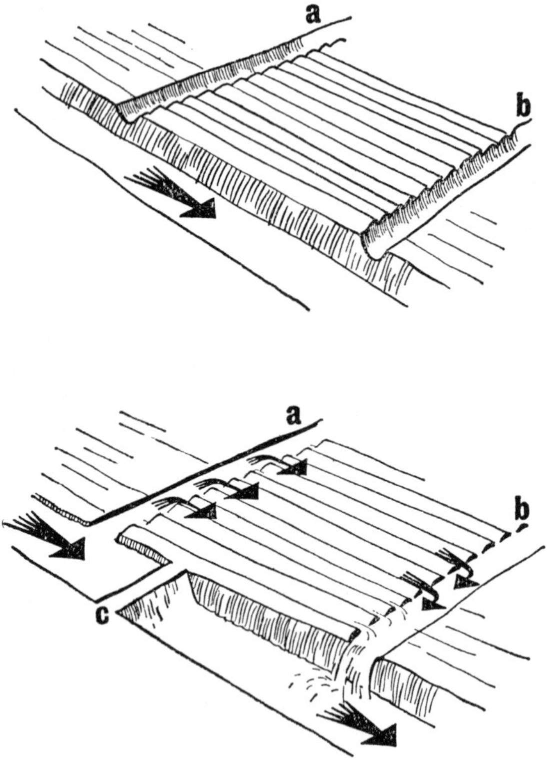

Fig. 29. Simple irrigation. The irrigation ditch is shown at (*a*) and the drainag ditch at (*b*). A weir across the stream at (*c*) causes the water level to rise c the upstream side and fill the irrigation ditch. The flow of water is show by the arrows.

Poisonous chemicals to kill insect pests and diseases. Dusts or sprays may be used to kill insects and diseases on growing crops. Smokes and gasses are used to kill pests in the soil and in stored crops. All these chemical poisons are expensive but the savings in damage to a valuable crop may well repay the cost.

Chemicals used to control weeds. Weeds may be killed by chemical sprays but there is always the danger of harming crop plants. It is essential to use the right kind of weed-killing spray in a particular crop. On most small farms the use of sprays would be more expensive than hand weeding.

Irrigation. This is the controlled addition of water to crops, when the soil is too dry for high yields. It is possible only when there is a good supply of water nearby, but even in wet countries irrigation can increase yields.

Simple forms of irrigation can be carried out quite cheaply. The traditional method of rice irrigation is to build low ridges around the field to prevent the natural flood waters from escaping on the flat land bordering rivers. A simple dam across a stream may be used to make the water flood over nearby fields. The water can be spread further by digging irrigation canals (Fig. 29).

Modern methods of irrigation include the use of sprinklers. The water is applied like rain through a series of holes in a pipe or from a rotating sprinkler head. Costly pumping equipment is required and these methods will pay for themselves only on very valuable crops.

7 Management of Livestock

Aims of livestock management

Livestock management is concerned with earning high gross margins from livestock enterprises. The main variable costs of keeping livestock are feed costs. The gross margin is therefore the value of production minus these costs. Gross margins may be increased by increasing the production or yield or by reducing costs of feeds (Fig. 30).

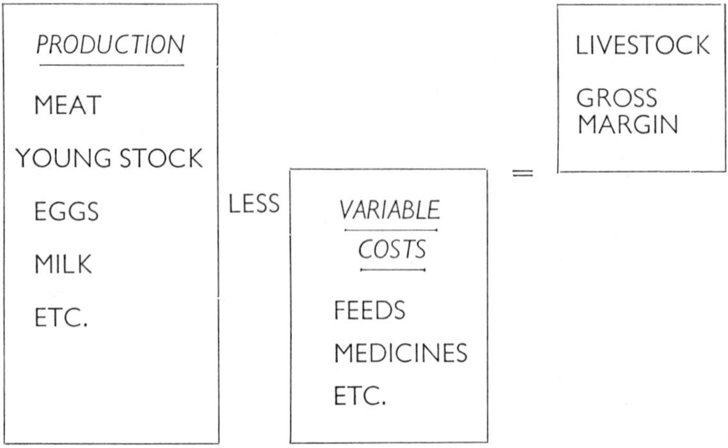

Fig. 30. Calculation of gross margin from livestock.

To maintain high gross margins the animals must be kept healthy. Medicines are expensive and the best way of keeping animals fit and well is by good management and feeding. Sometimes it is necessary to use medicines, dips and inoculations to control diseases and pests.

Livestock husbandry

To make a success of keeping livestock, a farmer must be a good livestock husbandman or must hire one to work for him. A good stockman understands his animals or birds and is able to handle them without making them frightened; a frightened or nervous animal does not produce well. He is able to recognize when stock are hungry, thirsty or sick so that they can be fed or watered or the illness treated before it becomes serious. A good stockman also knows when his animals are on heat and when they should be mated for breeding purposes.

These things cannot be learnt from books. To some extent stockmanship depends on a man's natural aptitude but it can be improved by experience. Before keeping livestock it may be worth while working with someone else's animals or poultry to get experience.

Livestock products

The importance of livestock products in human food was mentioned at the beginning of Chapter 4. Apart from meat, eggs and milk which are foods containing protein, livestock also produce hides and manure, and work in the case of draught animals.

Meat. Meat is muscle produced during the growth of animals. The value of animal production is the gain in value of the animal over a year. If the animal grows very slowly the yield or production is low. It can be increased by making the animal grow faster. Actual production can be measured either as the liveweight gain over a certain period or as the length of time necessary to grow the animal to slaughter weight.

Young stock. The production of a breeding herd or flock is measured by the number of live young produced in the year. This depends on the frequency of breeding, or how often the animal becomes pregnant. A cow that bears one calf in 2 years is less productive than one that bears a calf every year. The total number of young produced also depends on the number born and reared at each pregnancy. A sow that rears five pigs in a litter is less productive than one which rears ten.

Eggs. The total value of egg production will depend largely upon the number of eggs produced. In some countries eggs are graded by size and larger ones are worth more. The quality or grade of eggs can also affect the total value of production.

Milk. The total value of production from a dairy cow depends upon the amount of milk produced, usually measured in litres. The amount of milk produced in a year is affected by the yield during one lactation and the calving interval, which is the length of time from one calving to the next. A cow may produce 1000 litres in one lactation but, if the calving interval is 2 years, the annual production is only 500 litres (1000 ÷ 2).

Work. The production of draught animals is work. The value of work cannot be measured directly but the cost of getting the work done is the same as the cost of keeping the animal. However, work animals are not considered as a separate enterprise so there is no need to calculate a gross margin. Draught animals may finally be slaughtered for meat.

By-products. Hides, hair, hooves, horns, feathers and manure are all by-products of livestock enterprises. They are not the main product. If they are sold the value may be added to the value of the main product when calculating gross margins. Hides may form an important export in some countries.

Specialization in livestock production

Livestock can be divided up into two main groups: those which can live on grass or other green leaves and those that cannot. The first group includes cattle, camels, horses, donkeys, goats and sheep, in fact all grazing and browsing stock. Their cost of production must include the cost of maintaining grazing for them to feed on. The specialist producers of grazing stock are the nomadic herdsmen of the dry savanna regions.

The second group includes pigs and poultry, which do not feed on grassland, but eat feeds usually based on cereal grains. These animals are often called intensive livestock because they can be kept on very little land and so gross margin per hectare can be very high. The cost of cereal and other feeds for pigs and poultry may also be very high. With poor management it is very easy to spend more on the feeds than the total production is worth. These stock must be well managed and records should always be kept. The specialist producers of intensive stock are often not farmers at all. Pigs and poultry may be kept in backyard; poultry in battery cages and pigs in intensive fattenin houses. Under these factory conditions no farm land is used and all the food is bought.

Advantages of mixed farming

When livestock are kept on a farm where crops are grown the system is known as 'mixed farming'. Such a system has many advantages over specialization in either crops or livestock.

Crops and livestock supplement each other in the use of land. Pigs and poultry do not require land so these enterprises can be added to an existing farm without increasing the acreage. Grazing stock do require land but very often grassland can be used in a rotation in place of unproductive bush fallow. If this is the case the acreage of arable crops may not be affected by introducing grassland with grazing stock into the rotation.

Waste crop products may be used for livestock feed. For example the remains of the groundnut plant may be used to feed livestock. This makes a good animal feed, which since it is not a human food, would otherwise be wasted.

Manure is returned to the soil. The animal manure can be spread on the land to increase soil organic matter and fertility.

As a result the total farm production is increased. The average gross margin per hectare is therefore increased (see Chapter 3).

Assistance for cultivation. Cattle on a mixed farm may often be used to draw implements for crop cultivation. Donkeys, horses and camels may also be used for draught.

Improved management of stock. On a mixed farm cattle are herded in small groups. This makes it easier to give closer personal attention to each animal than is possible with large nomadic herds. It is also easier to give veterinary attention and to avoid losses by theft.

Advantages of diversification. The advantages of having several different enterprises, which were discussed under *Arable crop rotation*, also apply to mixed farming. These advantages are that labour requirements are spread more evenly over the year, income is spread more evenly over the year and risk is reduced.

Importance of good management

There is a widely held view that the only way of improving livestock yields is to use improved breeds usually imported from other countries. This view is wrong. A great deal can be done to improve the production of local breeds simply by good management.

When improved breeds are introduced they usually do not yield well unless the standard of management is very good. Various aspects of livestock management will now be discussed.

Animal health

Control of insect pests and diseases is essential for healthy growth and production. Insects and diseases thrive in dirty, hot and damp conditions, so perhaps the most important method of control is cleanliness. All livestock houses should be kept clean and the manure should be removed regularly, preferably every day. Manure should be heaped with grass to keep it dry and prevent it being infested by flies.

Sick animals should be moved away from healthy ones so that the disease cannot be spread. Straw or any other material that has been near the sick animal should be burnt. Kitchen swill should be boiled before it is fed to animals, as this will kill any germs which might cause disease.

Animals which appear quite healthy may be carrying a disease. If they contact other animals the disease may spread. This means that stock that have been in contact with those that are sick should also be separated from the rest of the herd or flock as they may be carrying the disease. A new animal brought on to the farm should not be mixed with the others until a period of isolation has passed. This isolation period, which may be about a month, is necessary to allow any disease the animal may be carrying to develop and be seen. If the animal does not fall sick during the isolation period, it is assumed that it is not a disease carrier and that it can be mixed with the other stock on the farm.

Cleanliness and isolation of sick animals are methods of preventing disease from spreading. Generally speaking it is better to prevent disease rather than to allow it to spread and then to try to cure it. Many animal diseases are carried by insects or other pests. Killing the pest will prevent the spread of the sickness. Ticks which live on the skins of livestock spread many serious illnesses. To prevent their spread, animals should be treated regularly to kill these pests. Cattle, sheep and goats should be dipped in a bath of solution which kills the pests. Alternatively cattle can be sprayed with a similar solution. Poultry may be dusted with a suitable powder.

One of the major diseases carried by insects is the 'nagana' or 'surra' sickness which is carried by the tsetse fly. Unfortunately it is difficult to control this fly, although bush clearing will help. The presence of the fly and the disease it carries has prevented mixed farming from being practised in many parts of Africa. Certain smaller breeds of cattle, such as the Muturu or the N'Dama and certain dwarf sheep and goats, are quite resistant to this sickness. If other stock are introduced into tsetse fly areas they must be injected with a drug regularly to control the disease.

Vaccination or inoculation may be used against other livestock diseases. Poultry may be inoculated to control fowl pox, fowl cholera, Newcastle disease and fowl typhoid and cattle against rinderpest and anthrax. Inoculation should be done by the local government veterinary officer.

Worms. Sickness in animals is often caused by worms, which live inside the animal. Practically all animals have a few worms, but when there are many the animal falls sick, becomes thin and weak and may eventually die. Even where the sickness is not obvious, stock are healthier and grow faster or produce more if the worms can be controlled. Control is often possible simply by good management if the way the worms live is understood.

Worms living inside the animal lay eggs and these pass out with the dung or manure on to the ground. In the case of the so called 'round worm', the eggs hatch out after 3 or more days. Animals eating the grass swallow the newly hatched worms which continue to live in the digestive organs and lay eggs (Fig. 31). The eggs cannot hatch out inside an animal. If animals are moved every 3 days they cannot take in worms because the eggs will not have hatched on the ground. In this way the numbers of worms cannot build up. When the animals have been moved the land must be left for long enough for the worms to die before it is grazed again. In the dry season a few weeks may be long enough but in the wet season land should be left for 2 or 3 months. In practice it may not be possible to move animals after 3 days and yet leave land for up to 3 months.

Flat worms which attack the liver usually spend part of the life cycle in fresh-water snails. Sheep suffer more than other animals from flat worms in the liver. They should therefore be kept away from swampy or wet areas where the snail might live.

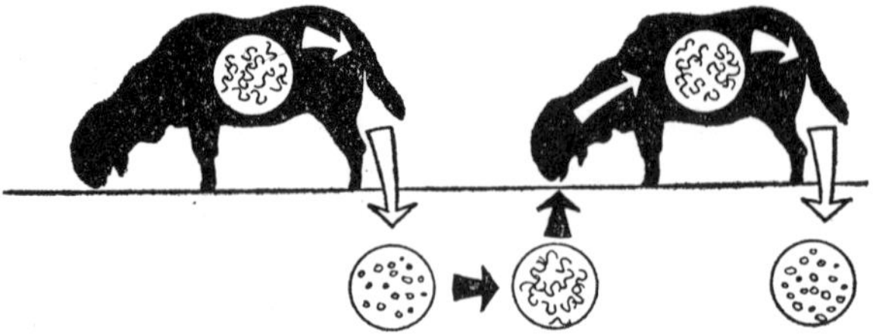

Fig. 31. The life cycle of round worms. Worm eggs are passed out with the dung or manure on to the ground. The eggs hatch out after a few days. Animals grazing the same land may take in these worms, which continue to live inside the animal and eventually lay eggs which are again passed out with the dung.

Tapeworms cause pork or beef 'measles'. These 'measles', which are round and white and about 1 inch across, are a stage in the life of the tapeworm. If pork or beef is not properly cooked the worm is still alive. A human eating the meat swallows it and the worm develops inside the human. The eggs pass out with the human manure. If this gets back on to the land pigs or beef cattle can swallow the eggs and the cycle begins again. To control these tapeworms human manure should not be spread directly on to pasture and pork and beef should be properly cooked.

Where careful management and change of grazing does not succeed in controlling worms, medicines must be used. Adult animals suffer less from worm attacks than do young stock and healthy well-fed animals suffer less than those that are not fit. Good feeding gives stock the strength to resist disease and worm attack.

The advice of the government veterinary officer should be sought if animals fall sick. He knows the right isolation periods for different stock, the right dips or sprays to use and how often to use them. He can help with inoculation of stock and with providing the medicines to kill worms.

Water supply

All animals and birds need drinking water in order to live. Before keeping any livestock on a large scale it is important to make sure that there is a reliable source of water throughout the year.

Feeding livestock

Animals must be fed a properly balanced diet or ration. Stock will not grow quickly or produce well if fed only on scraps and kitchen waste. A properly balanced ration must contain a mixture of different foods. Cereals alone cannot make a balanced ration. Different kinds of stock require different rations. Young stock, breeding stock, dairy cows and laying hens require more protein in the rations than do ordinary growing and fattening stock. For young stock, milk represents a balanced ration.

The livestock farmer must know when to wean young animals off their mothers' milk and on to solid foods. If this is done too early the health of the young may suffer. On the other hand it is important that the mother should have a rest period before having to start to feed new young ones, so weaning should be as early as possible.

Ready-mixed rations can be bought for intensively managed stock. These are more expensive than rations mixed on the farm but they may contain a better mixture of feeds and give better yields, particularly from poultry. A farmer who wishes to use home-grown food for his animals should get advice on the best mixture to use.

Feeds represent the biggest variable cost of livestock production. The gross margin depends upon the amount spent on feeds. However, if the extra production obtained by using a good ration is greater than the extra cost of using the ration, then the gross margin is increased. Food for intensive livestock is usually rationed according to the level of production or growth.

Grazing management

Grass represents a balanced diet for grazing animals. It is the cheapest form of livestock feed. However, for high production it should be well managed. Animals should not be allowed to graze freely over the whole grazing area. Free range helps the spread of disease and worms and allows the stock to eat only the best grass leaving the rest to waste. Heavy stocking on free range can lead to soil erosion. Animals should either be tethered and moved each day or grazed in paddocks. This is known as rotational grazing.

If stock are to be rotationally grazed in paddocks the grassland must be divided up by fences. If there are four paddocks and the stock are moved every 2 weeks each paddock is rested for 6 weeks before it is grazed again (Fig. 32). This rotation may not give complete control of worms and it is therefore necessary to give medicines regularly.

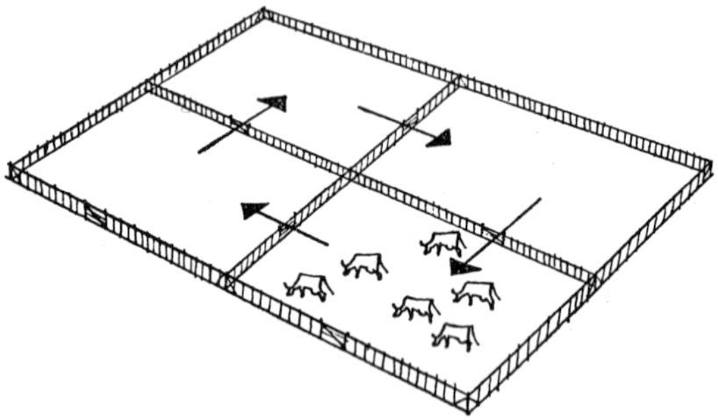

Fig. 32. Rotational grazing. The stock spends 2 weeks in each paddock before it is moved. After 2 weeks in the first paddock it is moved to the second paddock, and so on. It will graze the first paddock again after 6 weeks.

Rotational grazing not only controls the numbers of worms in the stock but also ensures that all the grass is eaten and rests the land between grazings. The condition of the soil and the grass is maintained or improved and the production of the grassland improved. Tethering or fencing stock prevents them from damaging growing crops.

During the dry season the amount of grass available falls and it is important to be prepared for this. Either stock numbers must be reduced or supplementary food is necessary. Grass may be preserved as hay or silage during the wet season as a supplementary food for the dry season.

Grazing animals need water and minerals. There should be a water supply and a mineral salt-lick in each grazing paddock. Tethered animals should be watered regularly and given minerals from time to time.

The production of a pasture may be measured in cow-days. It is necessary to record how many cows graze the pasture each day. The number of cows multiplied by the number of days gives the number of cow-days of grazing 'produced' by that pasture. Sheep and goats eat less than cows but, if the stock are mixed, sheep-days and goat-days may be converted into cow-days. If it is assumed that a cow eats four times as much as a sheep or a goat then the number of sheep-days or goat-days should be divided by four. For comparative purposes it is necessary to measure the size of pastures and calculate the cow-days per hectare.

Housing

All livestock in the tropics benefit from having some shade. Where there are no shade trees on grazing land it may be necessary to build shelters where the stock can go out of the sun during the heat of the day. Such shelters can also protect stock from beating rain.

Special housing is usually necessary for pigs, poultry and dairy cows, although for dairy cows the housing can be very simple. Carefully controlled management is necessary for these intensive stock, in order to get high gross margins. When they are housed it is easier to control the amount of food each animal is getting; there is less wastage of the expensive rations. It is easier and quicker to milk cows or collect eggs from stock gathered together in one building. It is also easier to check on the production of individual animals. Finally, clean production may be easier if the stock are housed. Manure can be removed regularly and the walls and floor kept scrubbed and clean.

Always get expert advice before putting up buildings for dairy cattle, pigs or poultry.

Breeding

All breeding animals and dairy cows must be mated regularly to maintain production. It is important to recognize when female animals are on heat and to get them served. Delays in getting an animal pregnant mean wasted time and loss of production. This is more difficult with cows since they are usually on heat about 5–10 hours at a time, unlike ewes, goats and sows which are on heat for 36 hours or

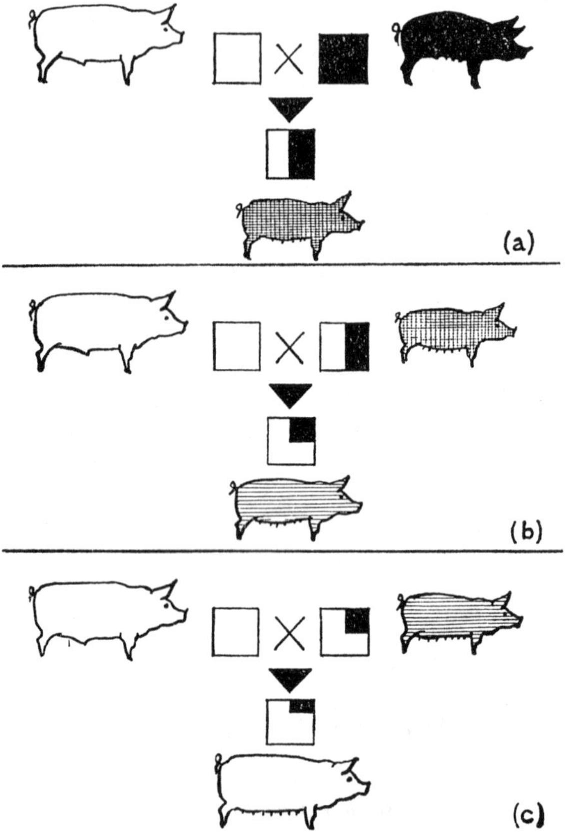

Fig. 33. Grading. The white pig on the left represents a boar of the new improved breed. The black pig on the right at the top represents a sow of the local breed. (*a*) If these two are mated, the young will all have half the characteristics from each parent. (*b*) If a half-breed sow is now mated with the improved boar, the young will all have three-quarters of their characteristics inherited from the improved breed. (c) If a three-quarter-breed sow is mated with the improved boar the young will be very similar to the new improved breed.

more. A few breeds of cattle, such as the East African Zebu, may have longer heats, but rarely more than 24 hours.

The sires and breeding females should be carefully selected from the most productive families. Generally it is advisable to buy new breeding males from time to time in order to improve the stock. If the farmer wants to introduce a new breed it may be cheaper and safer in the long run to use breeding males of this new breed. The process is known as grading and the way it works is illustrated in Fig. 33.

Many farmers buy all their laying poultry, dairy cows and fattening stock, so careful breeding on the farm is not necessary. But of course dairy cows need mating regularly in order to produce milk.

It is very important to know when to replace stock. With dairy cows or laying hens this should be when production falls to an uneconomic level. With growing stock, it should be when they reach the best weight for the market. These things may be very difficult to judge, although stockmen can attain the ability with experience. But for efficient livestock keeping it is better to keep production records. The number of eggs laid by each bird, the milk yield of each dairy animal and the weights of the growing stock should be recorded. These records are the surest way of knowing the right time to kill stock, besides helping the farmer to select stock for breeding and to estimate the amounts of food required.

8 Farm Records and Accounts

The advantages of record keeping

To manage his farm really well a farmer needs to keep some records and accounts. Simple record keeping need not take up too much of his time, and the advantages are numerous. Most other business men, even with very small businesses, consider it worth while to keep accounts. Some of them even hire an accountant to help with these records although their businesses are probably less complex than farming. The advantages of record keeping on the farm may be grouped under four headings.

1. *They give a history* of what has happened on the farm for the period during which they have been kept. By comparing 1 year's records with the next a farmer can see what progress he is making: whether his profits and yields are improving. It should also be possible to trace weaknesses in his farming methods.

2. *They are an aid to managerial control during production.* With the help of records a farmer can keep a close check on whether things are going according to his plans. He can check on whether he is using too much animal food or too much seed or whether crop and livestock yields are falling. It is important to detect where things are going wrong quickly so that they can be put right before big losses occur.

3. *They provide figures for farm planning and budgeting.* A farmer making plans to modify his farm needs to know what yields he can expect from crops and livestock and what costs and receipts he is likely to get. Many agricultural advisory officers keep approximate figures for the district but it is much better if each farmer has records of what has happened on his own farm.

4. *They tell a farmer how much he is earning.* With this information a farmer can make sure that he does not spend too much on himself and family. In this way he can avoid getting into money difficulties.

If a man feels that he has been charged too much income tax he may get a fairer assessment if he can present his accounts which show how much he earns.

Finally, a farmer has certain advantages in business dealings if he has accounts to show his earnings. For example, a bank is more likely to lend money to a man who can present well-kept accounts of how his money is spent.

Simple farm accounts

Keeping the account. Accounts are financial records of what the farmer spends and receives. Purchases and expenses should be recorded on one page and sales and receipts on the other. A simple account

1st Feb to 31st Jan							
SALES AND RECEIPTS				PURCHASES AND EXPENSES			
date	details	£	p	date	details	£	p
2/2	Yams - market		37	7/2	Bean seeds and sheep feed - J. Jones		45
3/2	Eggs - market		15				

Fig. 34. A simple account book.

book is illustrated in Fig. 34. This account was opened on the 1st February 1964. During the month of February farmer J. Udo sold nine yams for 38p on the 2nd, and twelve eggs for 15p on the 3rd. On the 7th he bought bean seeds and food for his sheep for 45p. These sales and purchases were recorded as shown in Fig. 34. Each time he bought or sold anything the farmer recorded the date, the details of what changed hands and the amount of money in his book.

On the 31st January 1965, Mr Udo closed his account for the year. He did this by adding up the figures in the columns to find out his total receipts and total expenses (Fig. 35). Of course, he could total these columns at any time but the usual accounting period is one

SALES AND RECEIPTS					PURCHASES AND EXPENSES				
Date		Details	£	p	Date		Details	£	p
2/2/	64	Yams - market		37	7/2/	64	Bean seed and sheep		
3/2		Eggs - market		15			Feed - J. Jones		45
10/2		Maize & eggs - market		20	12/2		Hoes - general store		50
19/2		Yams - M. Ali		15	14/2		Hired Labour - D. Peter	2	00
25/7		Eggs - market		15	17/2		Sheep Feed - J. Jones		95
23/9		Cocoa - produce buyer	11	59	20/2		Maize seed - market		11
15/10		Cocoa - produce buyer	18	78	10/3		Poultry Feed - Feed mill		70
2/11		Cocoa - produce buyer	14	12	10/6		Poultry Feed - feed mill		75
14/11		Sheep - B. Jones	1	90	15/7		Cocoa spray - general store	2	30
16/12		Hoes - B. Jones		40	22/7		Cocoa spray - general store	2	30
20/1/	65	Yams - market		50	29/7		Cocoa spray - general store	2	30
					10/9		Poultry Feed - Feed mill		63
					17/12		Sheep Feed - M. Ali	1	50
					19/12		Poultry Feed - Feed mill		65
					14/1/	65	Seed yams - B. Jones	2	59
					20/1		Sheep Feed - market	1	35
		TOTAL RECEIPTS	48	31			TOTAL EXPENSES	19	08

year. On the 1st February 1965 he opened a new account for the next year.

Produce used in the home. Mr Udo uses a large part of his production to feed the family. This part of production does not appear as sales in his account book. He therefore also keeps a record of what is used in the home. He writes down the quantity of a particular food that has been used and its estimated value. The value he records is equal to the price he would have had to pay if he had bought the food in the local market.

Date		Details	£	p
1/2	64	50 yams	1	23
10/2		15 yams		45
20/2		2 doz. eggs, 25 yams	1	00
7/3		1 sheep	1	75
15/3		2 doz. eggs, 15 yams		80
5/4		1 bag beans		35
19/4		2 doz. eggs, 20 yams		90
3/5		2 doz. eggs		30
17/5		30 yams, 1 basket vegetables	1	75
18/5		1 doz. eggs		15
1/6		2 doz. eggs, 1 old hen.		50
15/6		2 doz. eggs, 1 basket maize		91
11/7		1 doz. eggs, 1 old hen, 1 basket maize + vegetables	1	83
2/8		1 doz. eggs, 1 old hen.		35
16/8		2 doz. eggs		30
30/8		1 doz. eggs, 1 basket maize		65
10/9		2 doz. eggs, 1 old hen		50
21/12		1 doz. eggs, 1 old hen, 1 basket vegetables	1	05
5/1	65	2 doz. eggs		30
18/1		2 doz. eggs		30
		TOTAL VALUE	15	37

Fig. 36. Record of produce used in the home.

The record he kept from 1st February 1964 to the 31st January 1965 is shown in Fig. 36. He moved a lot of food at a time into the household store so he did not have very many entries in this record.

At the end of the year Mr Udo added up the total value of food used in the home. He added this to his total receipts to give the value of total production of the farm. By subtracting the total expenses he estimated his farm profit (Table 3).

TABLE 3

Estimate of Farm Profit for Mr J. Udo 1964–5

	£
Total receipts	48·31
Plus produce used in the home	15·37
Value of total production	63·68
Less total expenses	19·08
Farm profit	44·60

If he had not made allowance for the produce used in the home he would have assumed the value of his total production to be £48·31 and hence his farm profit as £48·31 less £19·08 which equals £29·23. He would therefore have seriously underestimated his production.

Comments on this method of accounting. The method of accounting just described is very simple and should not take up too much of a farmer's time. It would be suitable for any small farmer who wants to start record keeping. Books which are specially ruled for this purpose may be bought at any good stationery shop.

The estimate of farm profit would not be correct on a farm where big changes are being made in the farming system. For example, if a farmer is starting a tree crop plantation, the trees must be gaining in value (Fig. 2). He should include an estimate of this gain in value as part of the total farm production. Similarly where a farmer is building up a herd or flock the gain in value of the stock should be included as part of the total farm production.

If any valuable equipment is bought the total purchase price should not be included in the expenses of one single year. It is probably better to leave such items out of the account book. On the other hand the annual loss in value (depreciation) should be included as a cost each year (Chapters 2 and 5).

How these accounts can be used. Mr Udo had started keeping accounts in 1962, so he was able to compare his results for 1964–65 with those of the two previous years. These results are shown in Table 4. Clearly the total production of the farm had been steadily increasing. However, because the expenses were high in 1964–65 the farm profit fell. This farmer therefore decided to aim at keeping up the steady increase in production but to make special efforts to reduce his expenses.

TABLE 4

Comparison of Results for 3 Years

	1962–3	*1963–4*	*1964–5*
	£	£	£
Value of total production	50·52	59·68	63·68
Total expenses	14·00	14·50	19·08
Farm profit	36·52	45·18	44·60

The farm had been surveyed and was 5 hectares in area, so it was thus possible to work out the production, costs and profits per hectare. These figures are more suitable for comparison with other farms. Mr Udo has several neighbours who keep accounts so he is able to compare his results with theirs. His results for 1964–65 are compared with the average for similar farms in Table 5. Since men with similar farms were producing £20 per hectare against his £13 and since they earned £14 profit against his £9 he should be able to increase his profits if he increases his total production.

TABLE 5

Comparison of Production, Costs and Profit with Average

	Mr Udo's farm	*Average for similar farms*
	£	£
Value of production per hectare	13	20
Expenses per hectare	4	6
Profit per hectare	9	14

These are examples of the ways in which simple accounts can be used to help a farmer improve the management of his farm.

The accounts by themselves are not very useful in helping the farmer to find out which enterprises are not well managed. To do this he must keep records of individual crop and livestock enterprises.

Crop records

Know the size of your farm. In order to measure the productivity of the land and the success of farming operations it is essential to know the size of the farm. Productivity can be measured in terms of yield or profit per hectare. Yields of crops are useful only when related to the area grown.

It is therefore important to survey the farm and calculate the area. Surveying methods are described in Appendix 2. Preferably a plan should be drawn to scale showing all the plots and their areas. The business farmer should know the size of all the plots on his farm, even if he has not got a farm plan. The fallow land should also be measured as this is really part of the farm and will be used to grow crops at some time in the future.

Crop record book. A crop record should include details of the crops grown on each plot, the dates of planting and harvest, the amount of seed used and the yields. Notes may also be made of any treatments given. A suitable layout is given in Fig. 37. There should be a page for every plot on the farm but the page for plot 3 is the only one shown. A mixture of yams, maize and vegetables was grown in 1964–5 and the farmer has noted all the details. He has started the record for the following year with the planting of the cassava (manioc) crop.

Yields may be measured in the traditional way as so many bags, baskets or tubers. It is more accurate to record the actual weight harvested. To do this it is necessary to have a spring balance or scales. Apart from the advantages of accurate measurement of yields, many marketing boards and other large-scale buyers prefer to buy by weight.

If a weighing machine is available the weights of seed planted and of fertilizer applied can also be recorded.

The yield per hectare of a crop is found by dividing the total yield of the plot by the size in hectares. The total yields for plot 3 should be divided by $\frac{1}{3}$, which is the same as multiplying by 3. The yields per hectare from plot 3 are compared with the local averages in Table 6. The local averages are for the same crops grown in a mixture. The individual yields would be higher if each crop were grown alone (Fig. 12).

PLOT NO. : 3   NOTES : Some erosion in one corner.

HECTARES : 1/3

Crop	date of planting	amount of seed used	date harvest began	Yield	Notes (fertilizers, sprays, disease, etc.)
Yams	4/2/64	1 tonne	13/10/64	3 tonne	
Maize	27/2/64	4·5 kg	2/6/64	160 kg	Interplanted through the yams
Green Vegetables	15/3/64	2·5 kg	2/6/64	41 kg	Interplanted through the yams
Cassava	3/3/65	Old stem cuttings			

Fig. 37. A crop record book.

The yields from plot 3 are lower than average so the farmer should look for ways of improving his methods of crop production.

TABLE 6

Comparison of Crop Yields per Hectare (mixed crops)

Crop	*Farmer's plot 6*	*Local average*
Yams	9 tonnes	10 tonnes
Maize	160 kg	200 kg
Green veg.	41 kg	50 kg

Livestock records

Livestock numbers. Where livestock are kept these should also be recorded. Large scale pig keeping, poultry keeping and dairying require special records and it is not proposed to discuss these in detail here.

class of stock	no. on hand beginning	bought or received	born	total	slaughtered or died	sold or given away	no. in hand end
Sheep	2	–	3	5	1	1	3
Poultry	15	4	2	21	3	4	14

Fig. 38. Record of livestock numbers.

However, many farmers keep a few pigs, hens, goats, sheep or cattle for which records should be kept.

For all kinds of livestock it is important to keep a record of the number that are on the farm. This not only gives a check on thefts, deaths and losses but is also necessary to calculate yield per animal or per bird. An example of the layout for a livestock record book is given in Fig. 38.

A separate record should be kept for each class of stock. Records for sheep and poultry are shown. They may be completed weekly, monthly or annually, but they will be most reliable if kept every week.

At the end of each recording period the numbers can be checked. The total of the first three columns in the table should equal the total of the last three columns. If this second total is less, it means that some animals have been lost or stolen.

Production records. In the case of hens it is easy to record the number of eggs collected each day. These can be added up from time to time and divided by the number of laying hens to calculate the production in eggs per bird. Suppose a farmer records that he has collected 28 dozen (i.e. 336) eggs in 1 month. If he has fourteen hens the production per bird for the month must be 336/14 = twenty-four eggs. This amounts to almost one per day so the farmer can be well satisfied.

name or no. of ewe	served by	date served	date of birth of young	no. born	notes
No. 2	Ram from Min. of Agric.	9/1/64	12/6/64	2	Difficult birth

Fig. 39. A sheep breeding record.

The production of breeding stock consists of the young that are born. There are advantages in checking how many each breeding animal has produced. Those that are rearing many young can be used as parents of future breeding stock. Those that are reproducing badly can be removed from the breeding herd or flock. An example of how sheep births may be recorded is given in Fig. 39. Each time an animal becomes pregnant the date of service and the sire are recorded. When the young are born the date and the number born are written down. At the end of the year the total number born to each female can be worked out. The total of young reared on the farm divided by the number of breeding females gives the average production for the whole herd or flock. This figure may be a useful measure of efficiency for breeding stock.

The average production of laying hens or breeding stock, in fact of any class of stock, can be compared with local averages as a measure of efficiency in livestock production. A poor result suggests that the farmer should try to improve his methods.

Fattening records. For fattening stock the yield or production consists of growth or live-weight gains. The only satisfactory records for this kind of production are individual records. For each animal the dates of the beginning and end of the fattening period should be recorded. The length of the fattening period is a measure of the rate of growth.

However, where a balance or scales is available it is better to weigh growing stock regularly. If a farmer knows the weight of his stock, he can adjust the feed accordingly and can arrange to sell or slaughter at the right weight. Growth can be measured in terms of liveweight gain and if food records are kept this can be compared with the weight of food eaten. The ratio of the liveweight gain of growing stock to the weight of food is known as the 'conversion ratio' and is a useful measure of efficiency of meat production. For example, if a pen of pigs are fed 80 kg of feed and the total liveweight gain is 20 kg in a week, the conversion ratio for that week is 80/20 which equals 4·0.

General principles of record keeping

There are some general principles which apply to all record keeping. All records, to be of value, must be accurate, neat and full.

One way of making sure that records are accurate is by filling them in as soon as possible after the operation or transaction and by checking regularly. If possible records should be kept every day.

The other important rule in keeping accurate records is actually to measure quantities. It is no use guessing the area of land or yields of crops. Land should be measured with surveying tapes or chains and crop yields should be weighed.

The whole purpose of keeping records and accounts is to improve the standard of farm management. There is absolutely no value in spending time on records and calculations of profit, and production in individual enterprises, if no use is made of them.

All the results should be compared with some standards. The standards for comparison might be the results for previous years or the results for other farms. Government agricultural advisors may collect standards for comparison. Alternatively a group of farmers may decide to meet together from time to time to compare the results of their farming businesses. One farmer may find his management is better than that of his friends in one enterprise and worse in another. By comparing results and discussing problems together farmers can help each other to improve their management.

9 Records for the Larger Farm

The records already described should be sufficient for many small farmers. However, on a larger farm it is necessary to carry out farm valuations and to keep the accounts in more detail. Small farmers could benefit from keeping these more detailed accounts but they may not have sufficient time.

Where a farmer hires more than one or two workers he also needs to keep labour records. Valuations, detailed accounts and labour records are described in this chapter. Even on the large farm, records of crop and livestock production and of produce used in the home should be kept as already described.

Large-scale pig, poultry or dairy enterprises require detailed records but they are too specialized to be included in an introductory book such as this.

Farm valuations

To get a true estimate of farm profit, changes in the value of the farm should be included (Fig. 40). A farm should therefore be valued at the beginning and end of the accounting period. The value at the beginning is known as the opening valuation and the value at the end as the closing valuation. When one accounting period ends the next one begins immediately. The closing valuation for one period is therefore the same as the opening valuation for the next.

The *first step* in a farm valuation is to make a complete list of all livestock (animals and growing crops) and deadstock (machinery, buildings and stored products). It is a good idea to make the valuations and start the accounting period during the main dry season. At this time most of the food crops have already been harvested so there is no need to estimate their value as growing crops. The opening valuation of Mr Udo's farm which was made on the 1st February 1964, is given as an example in Table 7.

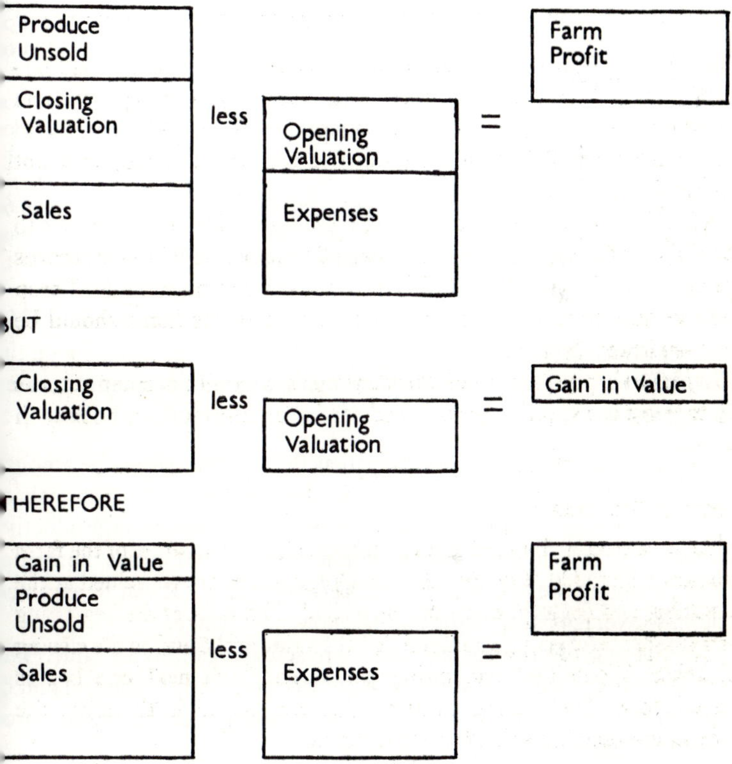

ig. 40. Calculation of farm profit. It can be calculated in two ways, either by adding the closing valuation to the sales and the opening valuation to the expenses, or by adding the gain in value to the sales. If the opening valuation had been larger than the closing valuation, then the farm would have fallen in value. The fall in value should be added to the expenses.

TABLE 7

Valuation of J. Udo's Farm, 1st February 1964

Item	*Quantity and price*	*Value* £
Livestock		
Sheep	2 at £2·00	4·0
Poultry	15 at 20p	3·0
Growing crops		
Cocoa	2 hectares of mature trees	75·0
Buildings	1 store	21·0
Machinery and implements		3·1
Stored products		
Yams	15 at 13p for 3	0·6
Maize	95 kg at 1p per kg	0·9
Beans	5 kg at 3p per kg	0·1
	TOTAL	107·9

The *next stage* in valuation is to estimate the value of each item on th list. It is very desirable that the same method of valuing should be use from year to year.

Livestock. Animals are best valued at the market price—the amoun for which they can be sold. With a breeding or dairy herd of animals o a laying flock of poultry the farmer will cull out certain unproductiv stock each year and replace them with new stock. In this case the farme is maintaining the overall value of the herd or flock. It can be assume that the average value of the animals does not change.

In the case of growing stock they are actually gaining in value. Th value must be estimated at the market price at the beginning and end o the year. The increase in value may be considered as the amount produced by the stock.

Machinery and implements. The value of a brand new machine o implement is the same as the cost. However, machinery immediatel starts to lose value, or depreciate. The value of a used machine is therefore the original cost minus the depreciation. Depreciation may b estimated roughly by assuming that a machine depreciates by $\frac{1}{5}$ o the value at the beginning of each year. A method of calculating th depreciation of all machinery and implements is illustrated in Table 8 The final figure obtained by subtracting the depreciation is the closin valuation for this year's accounts and the opening valuation for next year.

TABLE 8

Method of Calculating Depreciation of Machinery and Implements

	£
Opening valuation	3·15
Add purchases during the year	
2 hoes at 25p	0·50
Sub-total	3·65
Subtract sales during the year	
4 old hoes at 10p	0·40
TOTAL	3·25

Depreciation = $\frac{1}{5}$ × total = $\frac{1}{5}$ × 3·25 = 0·65
Closing valuation = 3·25 less 0·65 = 2·60

Buildings. The value of a building should be estimated in the same way as the value of a machine, although the building should last longer. It may be reasonable to assume a depreciation of 1/10 each year. Buildings worth £21 at the opening valuation would then depreciate by £2·10 and be worth £18·90 at the closing valuation.

Stored Crops. These should be valued at the market price on the valuation day.

Standing crops such as trees or cassava. Standing crops are rarely sold so they cannot be valued at the market price. In theory they should be valued at the cost of production while they are growing. This should include all the expenses involved in growing the crop. After trees have reached maturity they start to depreciate so the estimated value must be reduced each year by a suitable allowance. This depreciation allowance may be calculated by dividing the cost of production by an estimate of the useful life of the tree.

For example, if a palm tree costs £50 to produce and may be expected to go on yielding for 25 years, the depreciation allowance for each year should be £50 divided by 25. This equals £2 per year.

These calculations may become very complicated and it may be difficult to work out the cost of production. It may therefore be necessary to use some standard figures to estimate the value of growing crops. The local agricultural adviser may be able to help in suggesting suitable standard figures to use. If the farmer is constantly replanting to replace old trees the value of the plantation should not change. There is then no need to estimate the value of the plantation every year.

The cash analysis account book

This kind of account book is only slightly different from the simple account book already described. The only difference is that analysis columns are included so that the receipts and expenses can be listed under separate headings each time the farmer buys or sells anything. The cash analysis account book is illustrated in Fig. 41.

Each sale or purchase is entered twice, once in the total column and once in the analysis column, as illustrated. The sum of all the entries in the total column should always equal the sum of the entries in all the other columns. To illustrate the procedure the account from Fig. 35 has been entered in the cash analysis account book in Fig. 41.

Summarizing the account. At the end of the accounting period the cash analysis account book is summarized by finding the total of each column. The value of produce used in the home should be added to the total receipts. If possible the different items used should be separated and added to the respective columns as in Table 9.

TABLE 9

Total of Sales and Produce used in the House

	Total Receipts	*Yams*	*Maize*	*Other Food Crops*	*Cocoa*	*Sheep*	*Poultry*	*Hoes*
	£	£	£	£	£	£	£	£
Sales	48·31	1·02	0·05		44·49	1·90	0·45	0·40
Produce used in the house	15·37	4·49	1·87	2·51		1·75	4·75	
TOTAL	63·68	5·51	1·92	2·51	44·49	3·65	5·20	0·40

A closing valuation is also necessary to calculate the gain or loss in the value of the farm. This is completed in exactly the same way as the opening valuation. Table 10 shows the closing valuation for Mr Udo's farm, which was made on the 31st January 1965. The gain or loss in value of each item is also shown, gains being marked with a + and losses with a –.

Date	Details	Total £	Total p	Yams £	Yams p	Maize £	Maize p	Cocoa £	Cocoa p	Sheep £	Sheep p	Poultry £	Poultry p	Tools £	Tools p
1964															
2/2	Yams - market		38		38										
3/2	Eggs - market		15										15		
10/2	Maize & Eggs - market		21				6						15		
19/2	Yams - M. Ali		15		15										
25/7	Eggs - market		15										15		
23/9	Cocoa - produce buyer	11	59					11	59						
15/10	Cocoa - produce buyer	18	77					18	77						
2/11	Cocoa - produce buyer	14	12					14	12						
14/11	Sheep - B. Jones	1	90							1	90				
16/12	Hoes - B. Jones		40												40
1965															
20/1	Yams - market		50		50										
	Total receipts	48	32	1	03		6	44	48	1	90		45		40

Date	Details	Total £	Total p	Seeds £	Seeds p	Sheep £	Sheep p	Poultry £	Poultry p	Sprays £	Sprays p	Labour £	Labour p	Tools £	Tools p
1964															
7/2	Bean seed & Sheep feed - J. Jones		45		10		35								
12/2	Hoes - general stores		50												50
14/2	Hired Labour - D. Peter	2										2			
17/2	Sheep feed - J. Jones		95				95								
20/2	Maize seed - market		11		11										
10/3	Poultry feed - Feed mill		70						70						
10/6	Poultry feed - Feed mill		75						75						
15/7	Cocoa spray - general store	2	30							2	30				
22/7	Cocoa spray - general store	2	30							2	30				
29/7	Cocoa spray - general store	2	30							2	30				
10/9	Poultry feed - Feed mill		63						63						
17/12	Sheep feed - M. Ali	1	50			1	50								
19/12	Poultry feed - Feed mill		65						65						
1965															
14/1	Seed yams - B. Jones	2	59	2	59										
20/1	Sheep feed - market	1	35			1	35								
	Total expenses	19	08	2	80	4	15	2	73	6	90	2	00		50

Fig. 41. The cash analysis account book.

TABLE 10

Closing Valuation of J. Udo's farm at 31st January 1965

Item	*Closing valuation*	*Gain or loss in value*
	£	£
Livestock		
Sheep	6·00	+2·00
Poultry	3·00	no change
Growing crops		
Cocoa	74·75	−0·25
Buildings	18·90	−2·10
Machinery and implements	2·60	−0·55
Stored products		
Yams	0·88	+0·23
Maize	1·01	+0·06
Beans	0·15	no change
TOTAL	107·29	−0·61

The profit and loss account

The results of the year's farming as a business can now be summarized in a profit and loss account. As the name suggests this account gives a measure of the profit or loss obtained during the accounting period.

Expenses are listed on the left-hand side and receipts on the right The expenses include all sales together with the value of produce used in the home and the closing valuation. It is usual to make the two totals equal by including the profit or loss. If the farmer has made a profit, this is added to the left-hand expenses side. If the farmer has made a loss, this is added to the right-hand receipts side so that the two totals

TABLE 11

Profit and Loss Account for J. Udo's farm for the Period 1st February 1964–31st January 1965

Expenses	£	*Receipts*	£
Purchases	19·08	Sales	48·31
Opening valuation	107·90	Produce used in the home	15·37
Profit	43·99	Closing valuation	107·29
TOTAL	170·97	TOTAL	170·97

Note. The profit is added to the expenses side so that the two totals are equal.

are equal (see Table 11). It will be noticed that the farm profit estimated in this way is slightly different from the estimate given in Table 3. The result is more accurate when allowance is made for valuation changes.

The profit and loss account may be stated in a different way. The actual valuations may be omitted and the gain in value of trees and animals added to the receipts side and depreciation added to the expenses side (Table 12). Set out in this way the profit and loss account gives a true measure of total production, this being the total of the receipts side.

TABLE 12

Profit and Loss Account for J. Udo's Farm Showing True Value of Production and Costs 1st February 1964–31st January 1965

Expenses	£	*Receipts*	£
Purchases (excluding capital)	18·58	Sales (excluding capital)	47·91
		Produce used in the home	15·37
Depreciation		Gain in value of	
Cocoa	0·25	Livestock	2·00
Buildings	2·10	Stored products	0·29
Machinery	0·65		
Profit	43·99		
TOTAL	65·57	TOTAL	65·57

Summary

To summarize this method of accounting a brief outline is given below.

1. *Make an opening valuation.* First make a list or inventory of all live- and deadstock on the farm. Then estimate the value of each item at either the sale value or the cost of production.

2. *Keep a cash analysis accounts book.* Enter receipts and payments through the year, once in the total column and once in the appropriate analysis column. Total all columns at the end of the year. Check that nothing has been missed, by comparing the total column with the sum of the entries in the other columns.

3. *Keep a record of produce used in the home.* Add the value of this produce for the year to the receipts side of the account book.

4. *Make a closing valuation* on the last day of the accounting period. The procedure is exactly the same as for the opening valuation.

5. *Complete a profit and loss account.* Expenses, including all purchases and the opening valuation, are listed on the left-hand side. Receipts, produce used in the home and the closing valuation, are listed on the right. The difference between the two totals represents the profit or loss.

Gross margins

Accounts, if kept in sufficient detail, can be used to calculate the gross margins of individual enterprises. For this purpose the value of production and the variable costs of the farm must be divided between the various enterprises. The accounts must be planned with this in mind and each enterprise should have its own analysis column in the cash analysis account book.

If any of the farm crops are fed directly to livestock, some allowance must be made in calculating the gross margins. The value of this feed must be estimated and added to the value of production of the crops. The value of the feed must also be added to the variable costs of the livestock enterprises.

The gross margins for the enterprises on Mr Udo's farm are calculated in Table 13. As shown in Frontispiece the total farm gross margin less the fixed costs equals the farm profit (Table 14). The farm covers 5 hectares: 1·5 hectares of cocoa and 3·5 hectares of food crops and fallow in rotation. The gross margin per hectare for cocoa is therefore £37·34 ÷ 1·5 = £24·89. The gross margin per hectare for food crops is £7·43 ÷ 3·5 = £2·12. If the farmer wants to make good use of his land and earn high gross margins per hectare he should concentrate on cocoa rather than on food crops.

The sheep and poultry forage for their food and require hardly any productive land. The gross margin per hectare for the sheep and poultry is therefore very high. The average gross margin per hectare of the farm could be increased by keeping more sheep and poultry.

If the local average gross margins are available for comparison the farmer can check on his efficiency of production in each enterprise. To

TABLE 13

Calculation of Gross Margins on J. Udo's Farm for the Period 1st February 1964–31st January 1965

Variable costs	£	*Value of production*	£
Cocoa		Production approx.	
Depreciation of trees	0·25	200 kg at 22p	44·49
Spraying 1·5 hectares at £4·60	6·90		
GROSS MARGIN	37·34		
TOTAL	44·49	TOTAL	44·49
Arable crops		Production: Yams	5·74
Seeds: Yams	2·59	Maize	1·98
Others	0·21	Cowpea	0·70
		Cassava	1·21
GROSS MARGIN	7·43	Other crops	0·60
TOTAL	10·23	TOTAL	10·23
Sheep			
Feed costs	4·15	Production 2 old ewes	3·65
GROSS MARGIN	1·50	Increased valuation	2·00
TOTAL	5·65	TOTAL	5·65
Poultry			
Feed costs	2·73	Production 5 old birds at 20p	1·00
GROSS MARGIN	2·47	28 doz. eggs at 15p	4·20
TOTAL	5·20	TOTAL	5·20

TABLE 14

Profit and Loss Account of J. Udo's Farm showing Gross Margins 1st February 1964–31st January 1965

Fixed costs	£	*Gross margins*	£
Depreciation		Cocoa	37·34
Implements	0·65	Arable crops	7·43
Buildings	2·10	Sheep	1·50
Hired Labour	2·00	Poultry	2·47
FARM PROFIT	43·99		
TOTAL	48·74	TOTAL	48·74

calculate his gross margins per man day the farmer would have to keep labour records, as shown in the following section.

Labour records

Wages book. If the farmer hires labour it is a good idea to keep a wages book. An example is shown in Fig. 42. This is a record of the wages paid and is very useful if there is any argument with the workers about

workers name	hours worked	wage	extras (bonus or overtime)	total wage	value of food or drink given in part payment	actually paid
week beginning: 4/2/1964		£ p	£ p	£ p	£ p	£ p
Ignatius	40	1·50	–	1·50	0·10	1·40
John	40	1·00	0·30	1·30	0·18	1·12

Fig. 42. Wages book.

the wages they have received. When workers are given food or drink as part of their payment, its value should be added to the wages. This procedure has the advantage of making the workers realize what their total wages are.

Worker's name: Ignatius			Week beginning: 4/2/64	
day	plot no.	enterprise	work done	hours
Monday	3	Yam	Planting	7
Tuesday	5	Vegetables	Harvesting	3
	3	Yam	Planting	4
Wednesday	4	Maize	Clearing land	7

Fig. 43. Weekly time sheet.

Weekly time sheets. When calculating the wages of the workers it often helps to have a record of the hours worked. At the same time records of the hours worked on each enterprise are necessary for calculations of the gross margin per man-day.

A record of hours worked, known as a time sheet, is shown in Fig. 43. Time sheets may be filled in by the farmer or by the workers themselves. Another advantage of these records is that they give the farmer a check on what work his staff are doing.

Records must be planned according to the individual farmer's needs. A few examples have been given but many farmers will wish to plan their own records which would be more suitable for their own farms.

10 Farm Planning and Budgeting

The need for planning

The farmer who keeps records and accounts will be able to trace weaknesses in his farming methods. He should then try to correct these weaknesses. It may be that the farm system is satisfactory, but the day-to-day management is poor. This may result in low yields from crops and livestock and high costs for items like hired labour. If this is the case the farmer must try to improve his production techniques, without changing his farm system.

Alternatively it may be that the farm system is weak and should be changed. The farmer may be keeping pigs when he should be keeping poultry, or growing rice when he should concentrate on cotton. Before making a change it is a good idea to estimate what the results will be. It can only be an *estimate* of what the results will be, because we cannot see into the future. However, it would be foolish to change the farm system without having any idea of what the result will be. The method of estimating the results of a farming plan is known as budgeting.

The advantages of budgeting

Most farmers make some attempt at budgeting their farm plans, even though they may not work out their calculations on paper. It is well worth while to develop the habit of making more formal budgets. By doing so it is less likely that some important factor will be forgotten. Budgets can be used to decide whether a proposed plan will effectively increase profits. They can also be used by the farmer in deciding between two or more alternative plans. Once a plan has been decided upon, the budget can be used as a guide for production control. The estimated sales and costs can be used as targets against which the actual production can constantly be checked as the plan comes into operation. Management faults can be corrected before they have gone too far.

For a new farm, or a large-scale change in an existing farm system, a complete budget is necessary. For smaller changes in the farm system only variable costs are affected and a partial budget will be sufficient guide.

Planning procedure for partial budgets

Whether a partial or a complete budget is used the planning procedure is basically the same. There are four main stages which should be followed.

1. *List the available resources.* Any proposed farm plan must fit in with the available land, labour and capital, and with the farmer's ability as a manager. It is no good trying to make a change which requires more of these resources than the farmer can acquire.

The area of land available should be worked out. For any change in the cropping system it is preferable to have a map of the farm with the present crops marked on it and a record of the soil type and conditions. The yields of previous crops in past years may give an idea of the quality of the land. Otherwise it may be necessary to get the advice of the government agricultural officer who may take samples of the soil for analysis. This record of the available land will serve as a guide as to what crops are suitable and what hectares may be grown. It will also suggest what yields to expect.

Next, a list must be made of the available labour in man-days. This will include an estimate of the amount of labour which can be hired and the time the farmer and members of his family can give. It may be useful to show how much labour will be available in each month.

Available capital should also be noted. This will include the farmer's own savings, together with an estimate of the amount he can borrow if credit is available.

Finally, the farmer must take stock of himself as a manager. He must decide whether he would prefer to work with certain crops or with livestock. He must decide whether he would mind the extra efforts involved in the careful management of a crop like coffee or intensive livestock like pigs or poultry.

The resources available set a limit on the plans which are possible. Usually only one of these resources is a major limitation on what can be produced and the profit which can be earned. It is important to decide which resource is likely to be the major limitation.

2. *Decide which plan is possible.* There may be a whole range of alternative plans the farmer should consider. Many of these can be rejected immediately because the farmer is not interested in them or because he feels that he could not manage them properly. Other plans may be rejected because there is not sufficient land or labour or capital available to carry them out.

3. *Budget the possible alternatives.* At this stage estimates are made of the variable costs and the value of production for each of the possible alternative plans. These estimates are then used to calculate the gross margins. The gross margin for a particular plan is a measure of the amount that plan is likely to add to the farm profit.

Next, it is necessary to estimate or budget, for each plan, the requirements of the limiting resource. The plans can then be compared on the basis of the gross margin per unit of limited resource. If land is limited the plan giving the highest gross margin per hectare is best. If labour is limited the plan giving the highest gross margin per man-day is the best; and if capital is limited the plan giving the highest gross margin per £100 of capital is the best.

4. *Put the best plan into operation.* Once the best plan has been selected from the budgets, it must then be put into operation. This may take quite a long time, since tree crops and certain livestock do not reach full production for several years. New working methods and skills may be necessary.

Once the new plan is fully established it should run smoothly without too many management problems. The period during which the plan is put into operation is usually the most difficult and requires very careful management.

A partial budget where land is the limiting resource

As a simple example of a partial budget we will consider some farmer planning how to use 1·2 hectare of land. He has plenty of labour and sufficient capital for the plans he has in mind. He therefore wants to earn a high gross margin per hectare.

He has finally decided on two suitable alternative plans, either to grow rice or to grow cotton. His budgets for these two alternatives might be as shown in Table 15. At the estimated yields, prices and costs for the two crops, rice has the highest gross margin per hectare. The farmer would make best use of his limited supply of 1·2 hectare of land by growing rice.

TABLE 15

Partial Budgets to Compare Gross Margin from 1·2 hectares of Rice with Gross Margin from 1·2 hectares of Cotton

RICE

	£
Extra production from 1·2 hectares	
Sales 1·2 × 2000 kg milled rice at 3p per kg	72·00
Less extra costs for 1·2 hectares	
Purchase of seed 1·2 × 50 kg at 3p per kg	1·80
No fertilizer or sprays	—
GROSS MARGIN on 1·2 hectares per year	70·20
Gross Margin per hectare per year: £58·50	

COTTON

Extra production from 1·2 hectares	
Sales 1·2 × 1200 kg seed cotton at 5p per kg	72·00
Less extra costs for 1·2 hectares	
Purchase of seed 1·2 × 12 kg at 7p per kg	1·01
Fertilizer 9 bags superphosphate at 25p per bag	2·25
9 bags sulphate of ammonia at 30p per bag	2·70
Sprayer depreciation	0·25
Insecticide 1·2 × 35 litres at 26p per litre	10·92
GROSS MARGIN on 1·2 hectares per year	54·87
Gross Margin per hectare per year: £45·70 approx.	

If a resource other than land is limited, rice production may not be the best alternative.

A partial budget where labour is the limiting resource

If labour rather than land is the limiting resource then the man-day requirements of each plan must also be estimated (one man-day is usually assumed to equal 8 man-hours). The plan which is likely to give the highest gross margin per man-day should be chosen.

Suppose the farmer from the previous example had plenty of land but only 200 man-days of labour available for his change of plan. If he is choosing between growing rice or cotton he must estimate the labour requirements of each (see Table 16). At the gross margins already estimated and these labour requirements, cotton would give the highest gross margin per man-day. The farmer would make best use of his limited amount of 200 man-days of labour by using it to grow cotton.

TABLE 16

Partial Budgets to compare Labour Requirements and Gross Margin per Man-Day for Rice and Cotton

RICE

Labour requirements		Man-days per hectare
Cultivation and planting		100
Weeding		50
Harvesting and threshing		88
Total man-days per hectare		238
Gross margin per hectare (see Table 15)	£58·50	
Gross margin per man-day £58·50/238 =	£0·25	
Total gross margin which may be obtained 200 × 25p =	£50	

$$\text{Hectares required} = \frac{200}{238} = 0{\cdot}84 \text{ hectares.}$$

COTTON

Labour requirements		Man-days per hectare
Cultivation and planting		55
Weeding		30
Spraying		7
Harvesting		42
Clearing		2
Total man-days per hectare		136
Gross margin per hectare (see Table 15)	£45·70	
Gross margin per man-day £45·70/136 =	£0·34	
Total gross margin which may be obtained 200 × 34p =	£68	

$$\text{Hectares required} = \frac{200}{136} = 1{\cdot}47 \text{ hectares.}$$

This plan would be impossible if only 1·2 hectares were available. In that case only 1·2 × £45·70 = £54·84 gross margin could be made. Only 1·2 × 136 = 163 man-days would be used, so 37 man-days would be available for other purposes.

A partial budget where capital is the limiting resource

Where capital is the limiting resource the farmer would select the plan which is likely to return the highest gross margin per £100 capital.

As an example we will assume that a farmer has £200 capital to invest. He has sufficient land and labour to expand his business. He has finally decided that the capital should be spent to start either pig or poultry keeping. Pigs would be kept and fattened for bacon. To keep the plan simple the farmer would not do any pig breeding but would buy weaners as he needed them. Poultry would be kept for egg production and purchased as point-of-lay pullets. The budgets for these plans are shown in Table 17.

TABLE 17

Partial Budget to Compare Gross Margins per £100 Capital from Pigs and Poultry

PIG PRODUCTION

Basis

Purchase of a litter of 6 weaners and fattening to 95 kg liveweight every 6 months (Total of 12 fattened per year)

Extra production per year		£
Sales of 12 pigs at 95 kg at 14p per kg		159·60
Less extra costs per year		
Purchase of 12 weaners at £3·50		42·00
Feed costs of 12 pigs, 40 kg per pig at 15p* average price		72·00
Marketing costs 5% of value of production		7·98
	TOTAL	121·98

Gross margin £159·60 less £121·98 = £38 approx.

Capital required

Sufficient money to raise the first 6 pigs.
The money obtained from the sales of these pigs can be used to pay the expenses for the next litter to be fattened, and so on

	£
Building—cost of erecting accommodation for six fatteners	80·00
Purchase of 6 weaners at £3·50 each	21·00
Feed costs of 6 pigs	36·00
TOTAL CAPITAL	137·00

* N.B. The price of food per kg may appear very low. However the ration consists of a mixture of cassava at 3p per kg and a concentrate supplement at 27p per kg. The average price therefore works out at 15p.

$$\text{Gross margin per £100 capital } \frac{35}{137} \times 100 = £27{\cdot}74$$

If the farmer has £200 capital to spend he can keep $\frac{200}{137} \times 12 = 17$ pigs.

They should give a gross margin of $17 \times \frac{38}{12} = £53{\cdot}83$.

EGG PRODUCTION

Basis

Laying birds in battery accommodation. Unit of 48 layers.
Point of lay pullets to be purchased.

		£
Extra production		
180 eggs per bird from 48 layers at 18p per doz.		129·60
48 cull birds at 25p each		12·00
	TOTAL	£141·60
Less Extra costs		
Feed for 48 birds, 5 kg per bird at 32p		76·80
Replacements: 48 birds at 50p		24·00
Marketing costs: 5% of production		7·08
	TOTAL	£107·88

Gross margin: £34 approx.

Capital required

Sufficient money to erect housing and to purchase the birds
Feeds and other costs can be paid out of income from sales of eggs which should begin almost immediately

		£
Building—cost of erection		45·00
Battery cages 24 at £1·25		30·00
Purchase of 48 point of lay pullets at 50p		24·00
	TOTAL	£99·00

Gross margin per £100 capital: £34 on £99 = £34 approx.

If the farmer has £200 capital to spend he can keep two such units (96 birds altogether). This should give a gross margin of £34 × 2 = £68.

According to this particular farmer's estimates egg production will give a higher gross margin per £100 capital than pig fattening. He would therefore make best use of his £200 available capital in poultry keeping.

There may be other considerations which will affect the farmer's choice. For example, he may feel that the risks in egg production are greater than those in pig production. He may feel that he prefers to keep poultry. These personal considerations must not be forgotten in making the final choice. The budget helps in giving an idea of which plan is likely to make best use of the available land, labour or capital.

Planning the whole farm

The same principles apply when planning the whole farm except that the fixed costs must also be taken into consideration. For maximum profit, the enterprises which give the highest gross margins per unit of limiting resource should be chosen. However, it may be better to produce more than one kind of crop or animal. The advantages of growing different crops in a rotation and of mixed farming have already been discussed. Risks are reduced by producing several different kinds of products. Supplementary enterprises will increase farm profits.

On the other hand a farmer with a single enterprise may become a specialist in that kind of production. He may run the enterprise on a very large scale and become a very efficient manager. Probably the best solution for most farmers is to try to manage a few enterprises really well. In this way the worst dangers of monoculture and of too much diversification are avoided.

Planning an arable rotation

For the farmer who is going to grow arable crops it is necessary to plan a rotation. First of all he must decide what sequence of crops to grow. This decides the area of each crop. A 10-year rotation will require ten equal blocks of land. One of the crops in the rotation will be grown on each block.

As an example of planning we will take the following four-course rotation.

First year: Yam inter-planted with vegetables.
Second year: Early maize (fertilized) inter-planted with cowpeas and followed by late maize.
Third year: Cassava (manioc).
Fourth year: Cowpeas.

If a farmer is going to follow this rotation he must divide his land up into four equal-sized blocks (see Fig. 44). In the first year block 1 will be sown to yams, block 2 to maize, block 3 to cassava and block 4 to cowpeas. In the following year the crops will be changed. This is illustrated in the cropping programme in Table 18.

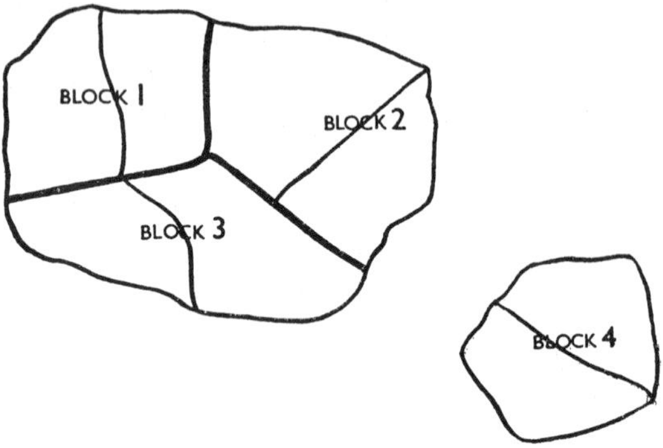

Fig. 44. Layout of blocks of land for rotation in Table 18.

TABLE 18

Cropping Programme for Farm shown in Fig. 44

	Block 1	*Block 2*	*Block 3*	*Block 4*
1st Year	Yam and vegetables	Early maize f.b. cassava	Cassava f.b. cowpeas	Early maize and sweet potato
2nd Year	Early maize followed by cassava	Cassava f.b. cowpeas	Early maize and sweet potato	Yam and vegetables
3rd Year	Cassava followed by cowpeas	Early maize and sweet potato	Yam and vegetables	Early maize f.b. cassava
4th Year	Early maize and sweet potato	Yam and vegetables	Early maize f.b. cassava	Cassava f.b. cowpeas

N.B. The crops on each block in the fifth year will be the same as those in the first year. The crops on each block in the sixth year will be the same as those in the second year, and so on. This arable rotation does not include any fallow. It is hoped that by using fertilizers and including the cowpea crops (which add nitrogen to the soil) the soil condition and fertility will be maintained. If this is not the case the rotation will have to be changed.

Complete budgeting

The choice of rotation decides how much of each crop will be grown. However, the farmer may wish to choose between several possible rotations. At the same time he may consider growing permanent tree crops or rice, or keeping livestock. He will probably possess several possible plans to choose from. To decide between different plans for the whole farm it is necessary to use complete budgeting.

In a complete budget an estimate is made of all the costs likely to be involved and of the total value of production. An estimate can then be made of what the profit will be. The plan which seems likely to produce the highest profit will be chosen if there are several plans.

An example. Budgets are often used to compare two or more different plans. They may also be used to estimate the outcome of a single plan. As an example of a complete budget a single plan for a whole farm is given in Table 19.

TABLE 19

Complete Budget for a Mixed Farm of 2 hectares Arable Rotation and Ninety-Six laying hens

			£
Gross Margins			
Arable rotation	Yams and vegetables	0·5 hectares	10·00
	Early and late maize and cowpea	0·5 hectares (fertilized)	14·00
	Cassava	0·5 hectares	6·00
	Cowpeas	0·5 hectares	1·00
96 laying hens			68·00
		TOTAL GROSS MARGIN	99·00
Fixed costs			
Labour—one regular hired worker			72·00
Depreciation of poultry housing £75 over 15 years			5·00
Hand tools and equipment			2·00
		TOTAL FIXED COST	79·00

Farm Profit: Total Gross Margin less Total Fixed Cost = £20

The plan is to maintain 2 hectares of arable land under the rotation given above and to keep a unit of ninety-six laying hens. The estimated farm profit from this plan is only £20 per year. There is a high fixed cost involved in hiring the regular worker. If the farmer could manage to run the farm without hiring any labour the profit would be increased by £72 to a total of £92.

It may be possible to improve on this plan by making some small change. The effect of the change on the farm profit can be estimated by preparing a partial budget.

Budgets for new equipment

A special case arises where a farmer thinks about buying a new piece of equipment. Since this represents only a relatively small change, a partial budget is suitable. The extra profit is the difference between the extra returns and the extra costs.

However, the buying of new equipment means that the fixed costs of the farm are changed. The extra costs must therefore include the extra fixed costs of depreciation, maintenance and repairs.

A budget for a grain storage bin

As an example we will take a farmer who is considering buying a concrete bin to store his guinea corn. He estimates that he will have 10 tonnes to store and that a bin of this capacity will cost £100. He expects the bin to last 10 years, so that annual depreciation will be £100÷10 = £10. There should be no maintenance or repair costs.

If the grain is treated to kill insects and the bin is sealed the losses due to insect damage and drying out should be very small. In fact these losses can be ignored.

The budget to estimate the extra profit is given in Table 20. If the return of £10 per £100 capital is higher than the interest which would be charged on this amount, then the plan to build the store is a good one. Of course, if there is a mistake in the budget such as an over-estimation of the extra returns or the underestimation of the extra costs, the plan may not be so good.

It should be noted that as the bin depreciates its capital value falls. As the capital value falls the return per £100 capital should get bigger.

TABLE 20

Partial Budget to Estimate the Profit from Storing Guinea-corn

	£
Extra production	
Sale value—10 tonnes at £24·25 = £242·50	
Value at harvest time—	
10 tonnes at £20·00 = £200·00	
Gain in value	42·50
Less extra costs	
Cost of fumigating grain at £0·15 per tonne	1·50
Depreciation of bin £$\frac{100}{10}$	10·00
Cost of cleaning bin	0·25
Extra profit on 10 tonnes	30·75
Extra profit per tonne	3·08
Capital needed in the first year	
To buy bin	100·00
To buy guinea-corn	200·00
	300·00

Return on capital in the first year: £30·75 on £300 capital = £10·25 per £100 capital

The effect of not using equipment to the full

If this farmer were able to store only 5 tonnes of grain in the bin the result would be very different. A budget using the same estimates that were used in Table 20, but having only 5 tonnes stored, is given in Table 21. In this case the average return on £100 capital is only £5 as compared with £10 when 10 tonnes are stored.

This lower figure obtains because of the effect of fixed and variable costs discussed in Chapter 2. The costs of bin depreciation and cleaning the bin are the same no matter how much grain is stored. When the bin is full these fixed costs apply to a large amount of grain, so the cost per tonne is low. When the bin is not full the fixed costs apply to a small amount of grain, so the cost per tonne is high.

This example illustrates a general principle that a piece of equipment or a building should be put to as much productive use as possible. In this way fixed costs are spread over a greater production, and profits are increased.

TABLE 21

The Effect of not using a Storage Bin to its Full Capacity

	£
Extra production	
Sale value—5 tonnes at £24·25 = £121·25	
Value at harvest time—5 tonnes at £20 = £100·00	
Gain in value	21·25
Less extra costs	
Cost of fumigating grain at 15p per tonne	0·75
Depreciation of bin £$\frac{100}{10}$	10·00
Cost of cleaning bin	0·25
Extra profit on 5 tonnes	10·25
Extra profit per tonne	2·05
Capital needed in the first year	
To buy bin	100·00
To buy guinea-corn	100·00
Total capital	200·00

Return on capital in the first year: £10·25 on £200 capital = £5 per £100 capital

A budget for machinery

The final illustration budget is for a plan to buy a tractor and implements. Such a plan might be proposed for a farmers' co-operative. A partial budget to compare extra returns and extra costs is suitable. In this case, however, the largest part of the extra returns is the saving in labour costs. The plan is to use a tractor and implements instead of temporary labour to cultivate and to plant 32 hectares of arable land. Maize and groundnuts are grown during the first rains and cotton and beans during the late rains. The tractor is expected to cost £700 and last 5 years and the implements £1,000 and to last 8 years. Machinery operation (fuel, oil and repairs) will cost 25p per hour and temporary hired labour 25p per man-day.

A simple budget for this plan is given in Table 22. According to this estimate it would result in a loss. This suggests that it would not be wise to carry on with the plan.

TABLE 22

Partial Budget for the Purchase of a Tractor and Implements

	£
Extra returns	
Value of labour saved	
Clearing, ridging and planting 32 hectares of first-rain crops at 60 man-days per hectare—1920 man-days at 25p	480·00
Ridging and planting 32 hectares of second-rain crops at 15 man-days per hectare—480 man-days at 25p	120·00
Total extra returns	600·00
Variable costs	
Ploughing, ridging and planting 32 hectares of first-rain crops at 12·5 hours per hectare—400 hours at 25p	100·00
Ridging and planting 32 hectares of second-rain crops at 7·5 hours per hectare—240 hours at 25p	60·00
Fixed costs	
Annual wage of driver-mechanic	192·00
Depreciation of tractor £$\frac{700}{5}$	140·00
Depreciation of implements £$\frac{1000}{8}$	125·00
Total extra costs	617·00
Difference: Total extra returns less total extra costs: Loss	17·00

On the other hand, if the assumption had been slightly different the budget might have shown a profit. If the tractor could be put to more productive use, the fixed costs would be spread over greater production and profits would be increased. This could be achieved by using the machinery on a greater area of land or by using it to save cost of labour at other times of the year. A tractor and implements may make better and speedier cultivations than can be done with hand tools. This could result in higher crop yields. If this is likely, an estimate of the extra crop yield should be added to the extra returns in the budget.

How to make the estimates

To prepare a budget, estimates are needed of the yields, prices, costs and requirements of land, labour and capital of the enterprises included. These estimates must be based on past experience.

On an existing farm, records and accounts will give the necessary figures. Plans can be based on these past results. For the man starting a new farm it may be possible to get help from the local agricultural advisory officers in deciding what to produce. Where this is not possible it is probably best to rely on the advice of experienced local farmers.

Appendix 1

Farmers' Co-operatives

Self-help through co-operation

Examples have been given, all through this book, of ways in which farmers can help themselves by co-operation. Farm layout can be improved if farmers will co-operate in a land consolidation scheme (Chapter 3). Roads, schools, health centres and village halls can be built by the people working together in communal labour (Chapter 4). Farmers can co-operate by meeting together to discuss their management problems and compare their records and accounts. In this way they help each other to improve the management of their farms (Chapter 8).

These are examples of neighbourly help which any farmer can try without needing a special organization. However, there is considerably more scope for farmers to help themselves by organizing themselves into a regular co-operative.

Co-operatives of one kind or another exist in practically every country of the world and bring considerable benefit to their members.

Kinds of co-operatives

Processing. Practically any kind of crop- or livestock-processing operation can be done by a co-operative. Examples of crops commonly processed in this way in tropical countries are coffee, rubber, oil palm and cassava. Livestock products such as milk which is processed into cheese or butter, and meat which is cut up and prepared for marketing, are sometimes handled and distributed by co-operatives.

These operations can be done more quickly, efficiently and cheaply by working as a group, and co-operatives may be able to afford better processing equipment than an individual farmer.

Marketing. This is frequently linked with processing in the same co-operative. Having done the processing, the same organization sells the product. In order to get a good price it may be necessary to store the product or to transport it to the buyer.

Storage and transport is almost always cheaper per article or unit of weight when large quantities are involved. Buyers, too, often prefer to grade and buy in large quantities. Co-operative marketing is therefore preferable to market bargaining by individual farmers.

Buying. A natural development from co-operative selling is co-operative buying. Materials such as seeds, sprays, fertilizers or tools may cost less if they are bought in large quantities. A buying co-operative can save its members money by buying in large quantities.

Machine sharing. One of the best ways in which small farmers can mechanize is by co-operating to buy a machine which would not be profitable on an individual farm. If the machine is used on several farms, its fixed costs are spread over several enterprises which give a large total production.

Co-operative services. In some parts of the world farmers' co-operatives provide other services. In the United States of America electricity is distributed by such organizations. In some countries the farmers co-operate to hire their own agricultural advisers. Improved bulls and other breeding animals may be owned by a co-operative and used by all members.

Almost any kind of service could be provided in this way so there is obvious scope for farmers to improve their incomes and living conditions by co-operation.

Co-operative banks. The provision of loans is a particular service which deserves separate mention. By putting their savings together members of a co-operative can provide a large sum of money. Out of this sum loans can be made to individual members as the need arises.

Governments and other organizations are usually willing to support co-operative banks by lending them money. This means that the banks are then able to help the members by making larger loans.

Co-operative farming. The closest kind of co-operation occurs where the farmers actually grow their crops and rear their livestock as a co-operative.

Co-operative farming can bring the benefits of all the other kinds of co-operation mentioned above. To be successful the members must be interested in working *for the common good* rather than for themselves as individuals.

Multi-purpose co-operatives. A single co-operative may combine several activities such as processing, marketing and other services including credit. These multi-purpose co-operatives may be particularly successful in developing countries as they unite the common interests of the farmers in a larger organisation.

Starting a co-operative

In all the kinds of co-operative described the basic idea is that the members help themselves and each other. They do not rely entirely upon the government or the services of an adviser or an administrator. On the other hand, it is advisable for the members to seek outside help in setting up and developing the organization. In many countries there is a national organization concerned with helping co-operatives which gives advice and sometimes lends money. When it is proposed to start a co-operative it is obviously worth trying to get advice and help from such a national organization.

The basic essentials for starting a co-operative are a group of farmers who want to work together and leaders to do the organizing.

A group of farmers who want to work together. The true co-operative spirit exists when people genuinely want to work together for the common good. On some occasions it may be necessary to put the interests of the organization above one's own interests. In some cases it may be necessary to share possessions or money with other members.

Without this desire to work together and to share, a co-operative is unlikely to succeed. If members are interested only in themselves, and form the organization mainly to get help from outside, they will never be able to work together happily and effectively.

Leaders. The success of a co-operative frequently depends upon the character and ability of the leaders. They must be men whom the others can trust and follow, and be men who will try to lead rather than control their fellow members.

Leaders are necessary, first of all, to interest other farmers in joining the co-operative. In spite of the benefits to be gained, certain farmers may need quite a lot of convincing, before they will join. Secondly, leaders are necessary to direct the policy of the co-operative. In a large co-operative it is usual for the members to elect a board of directors and a chairman. Finally, leaders are necessary to assist in the day-to-day management. Since farmers are busy with their own farm businesses they may need to hire someone to be a manager. But in a small co-operative the affairs of management may be performed by the secretary.

Some co-operative principles

There are certain principles which are followed in many co-operatives and which might serve as a guide to farmers who plan to start one. Some of the most important principles are as follows.

1. *Open membership.* Any person who is likely to benefit from the services of the co-operative should be free to join. On the other hand nobody can be compelled to join. In some cases it may be necessary to prevent a man of bad character from joining, but it would be harmful to keep people out for political, religious or tribal reasons.

2. *Roughly equal shares in the money contributed.* Much of the capital for developing a co-operative must be provided by the members themselves. It is preferable if each member contributes roughly the same amount of money. If one member provides a large part of the capital he may consider that he is entitled to control the running of the co-operative.

3. *Limited interest paid on capital.* The members of the co-operative may be paid interest on the capital they have invested, but only at a moderate fixed rate. The interest rate does not rise when the co-operative is prospering and fall when it is not.

4. *The money surplus is distributed in proportion to the amount the co-operative is used.* The prices charged by a co-operative for its services are just sufficient to cover the costs of running the organization, including interest on the capital. Nevertheless, if prices cover *all* costs there will usually be a money surplus. Some of this surplus may be used to buy new equipment for the co-operative, but any surplus not used in this way should be distributed to the members. It is distributed on the

basis of the amount of business each member has done with the co-operative and not on the basis of the amount of capital he has put into it.

5. *Equal voice in control.* In reaching decisions on the running of the co-operative the rule should be: one man, one vote. A particular member should not have a greater say simply because he has put more capital into or has done more business with the co-operative.

Other principles will apply in particular situations, and members will develop rules of their own as the society grows. However, at some stages of growth most co-operatives make some move towards the education of their members and staff.

Appendix 2
Surveying

Areas and dividing of land

Calculating areas. The area of a piece of land is measured by dividing it up into units and calculating the area of each unit. The unit used may be a triangle, which has three sides, or a trapezium, which has four sides two of which are parallel. These shapes are shown in Fig. 45.

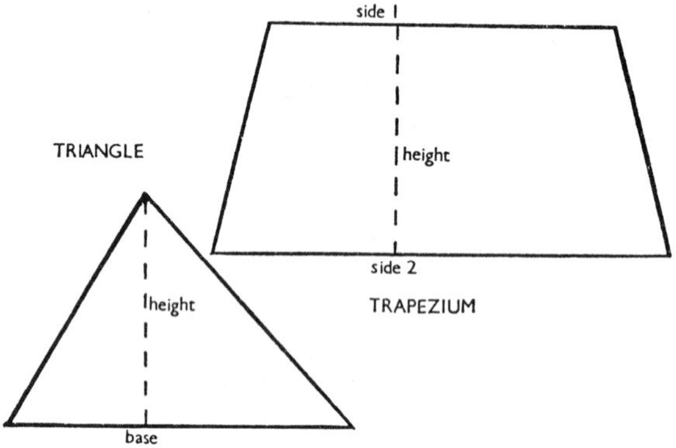

Fig. 45. Triangle and trapezium.

The area of a triangle is calculated by the formula

$$\text{area} = \frac{\text{base} \times \text{height}}{2}$$

For example, if the base of the triangle in Fig. 45 were 30 metres long and the height were 20 metres, the area would be

$$\frac{30 \text{ metres} \times 20 \text{ metres}}{2} = 300 \text{ square metres.}$$

The area of a trapezium is calculated by the formula

$$\text{area} = \frac{(\text{side 1} + \text{side 2}) \times \text{height}}{2}$$

For example, if the length of side 1 of the trapezium in Fig. 45 were 30 metres, of side 2, 40 metres, and if the height were 20 metres, the area would be

$$\frac{(30 \text{ metres} + 40 \text{ metres}) \times 20 \text{ metres}}{2} = 700 \text{ square metres.}$$

It will be noticed that the areas are estimated in *square* metres. Distances are measured in centimetres, metres and kilometres but areas are measured in square centimetres, square metres and square kilometres. A square metre is an area equal to that of a square with each side one metre in length.

The relationship between various measures of area are given in Table 23. Although there are 100 centimetres in one metre there are 100 × 100 or 10,000 square centimetres in one square metre.

TABLE 23

Table for Converting Square Centimetres, Metres, Kilometres and Hectares

1 square metre	=	10,000 square centimetres
1 are	=	100 square metres
1 hectare	=	10,000 square metres
1 square kilometre	=	100 hectares

Marking out the land in triangles. Fig. 46 shows a plot of land which is to be surveyed by dividing it up into triangles. There is no need to make lines on the ground provided that the corners are marked by poles stuck into the ground. The poles are placed so that the imaginary lines joining them would include as much of the plot as they would exclude.

The position of the poles are marked × in Fig. 46. The imaginary lines joining the poles are shown dotted and, since the boundaries of the plot are crooked, some of the plot lies outside and some inside the dotted lines. The area outside is approximately equal to the area inside.

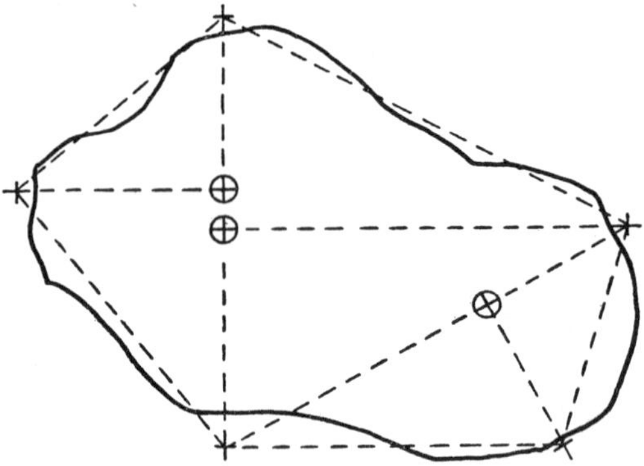

Fig. 46. Marking out a plot in triangles.

In order to calculate the area of the triangles it is necessary to measure the base and the height. To get an accurate result the height must be a right-angles or square to the base. The poles marked ⊕ in Fig. 46 were placed so as to form a right angle at that point.

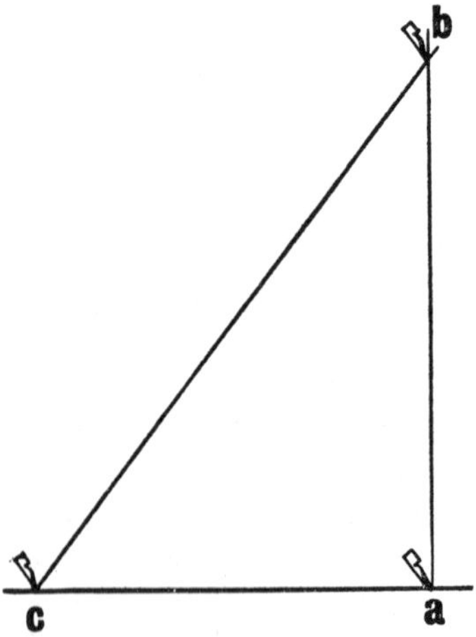

Fig. 47. Marking out a right angle.

Marking out a right angle. The method used depends upon the fact that if the sides of a triangle are in the ratio 3:4:5 the triangle is right-angled. From the point *a* in Fig. 47, where the right angle is to be marked, 3 metres are measured back to *c* and pegs are stuck into the ground at *a* and *c*. A piece of cord 4 metres long is tied to the pegs at *a* and one 5 metres long tied to the peg at *c*. These two cords are stretched out and another peg is stuck into the ground where they meet at *b*. The angle marked at *a* is now a right angle.

When marking out the heights of the triangles there is no way of knowing where point *a* the right angle, should be. The position must be guessed first of all and then a check must be made to see whether it is a right angle. If not, the pole must be moved and another check made. The final position of the poles marked ⊕ in Fig. 46 is thus found by trial and error.

Marking out in strips (each strip being a trapezium). Fig. 48 shows a plot of land which is to be surveyed by dividing it up into strips.

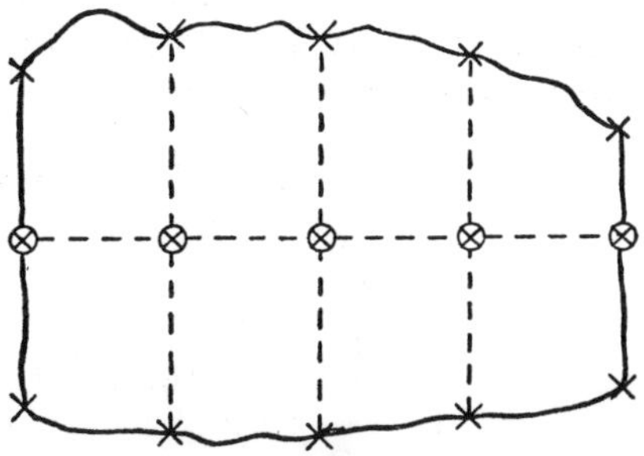

Fig. 48. Marking out a plot in strips.

Again, the marking out is done with poles stuck into the ground. The plot is divided up into several strips which are equal in width. The width is equivalent to the height of the trapezium shown in Fig. 45. The sides of each strip must be set out at right angles to the line on which the width of each strip was measured. The length of each side of each strip must also be measured to calculate the area.

Measuring distances

Pacing is a simple way in which one man can measure distance without any special equipment. All that is necessary is to count the number of steps or paces it takes to walk the distance. It is much better to walk along normally rather than to take big strides.

For very rough measurements 1 pace can be assumed to equal 1 metre but it is of course more accurate to measure the length of a pace. The best way to do this is: pace a measured distance of at least 200 metres three or four times and calculate the average number of paces taken. By dividing 200 metres by the average number of paces the length of one pace is found. Thus, if 200 metres took 220 paces, the length of one pace would be

$$\frac{200}{220} = 0{\cdot}909 \text{ metres.}$$

Any other distances paced out can be estimated in metres by multiplying the number of paces by this factor.

Tapes. More accurate measurements can be made by using a tape. Tapes may be made of cloth or steel and can be bought in most hardware stores. Tapes as shown in Fig. 49 are rolled up in a circular case

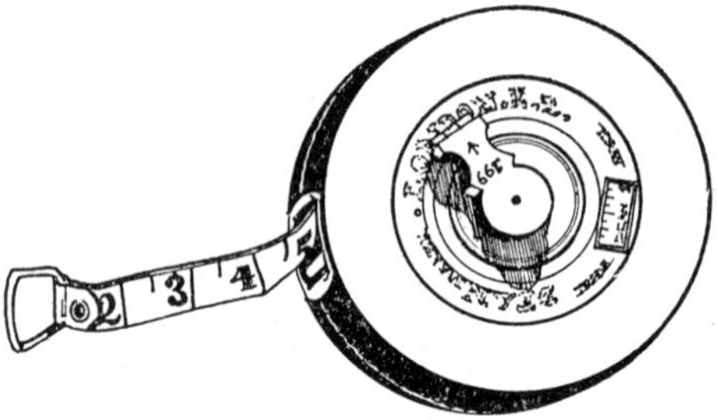

Fig. 49. A surveyor's tape.

when not in use. They are usually 20 or 30 metres long and are marked in centimetres. Tapes should be kept clean, and dry, and steel tapes should be wiped with an oily cloth after use.

When measurements are made the tape must be stretched out straight over the ground, so two people are needed: one to hold the end and the other to make the measurement. If the distance to be measured is longer than the tape, two measurements must obviously be made. The man doing the measurement sticks a peg in the ground or makes some mark at the end of the tape, and the tape is moved along and the second man holds the other end at the mark while the first man makes the second measurement (Fig. 50).

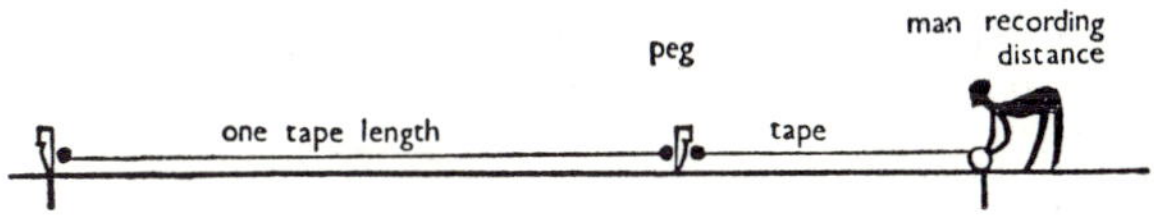

Fig. 50. Using a surveyor's tape

The distance between two points on sloping land is different from the distance on the flat (Fig. 51). There may therefore be errors in distances measured on slopes. However, these errors are not serious except on steep slopes, and may therefore usually be ignored when measuring farm plots.

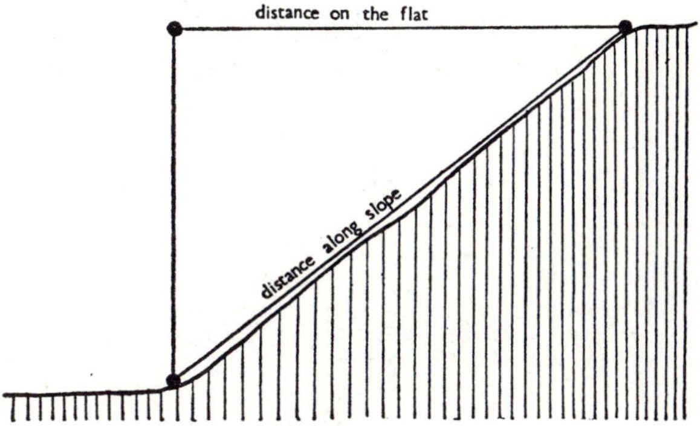

Fig. 51. The distance between two points on sloping land is different from the distance on the flat.

Examples of surveying

A survey by dividing land into triangles. The plot shown in Fig. 52 was marked out as shown, and the base and height of each triangle was measured by pacing. A former trial showed that 200 metres took 222 paces, so the length of one pace was 0·9 metres. All the distances measured in paces were multiplied by 0·9 to convert the result into metres.

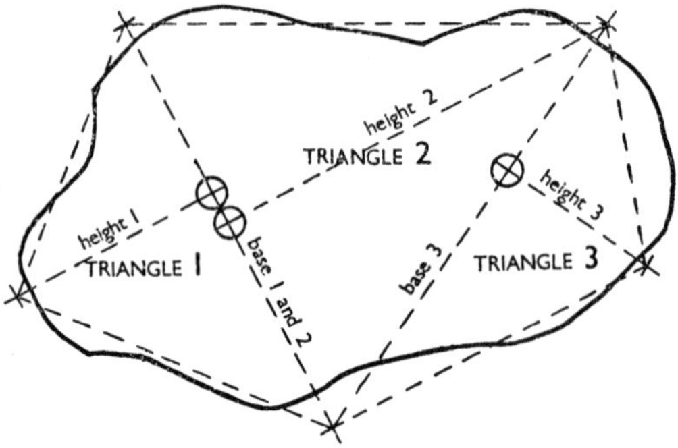

Fig. 52. Plot I.

The results are given in Table 24. The total area, 1297 square metres, can be converted into hectares by dividing by 10,000. The area of the plot is therefore 0·13 hectares.

TABLE 24

Results of the Survey of the Plot shown in Fig. 52

	Base		*Height*		*Area*
	Paces	*Metres*	*Paces*	*Metres*	*Square metres*
Triangle 1	39	35	30	27	$\frac{35\times 27}{2} = 472{\cdot}5$
Triangle 2	39	35	37	33	$\frac{35\times 33}{2} = 577{\cdot}5$
Triangle 3	42	38	14	13	$\frac{38\times 13}{2} = 247{\cdot}0$
TOTAL AREA					1297·0

A survey by dividing the land into strips. The plot shown in Fig. 53 was marked out and the two sides and the height of each strip were

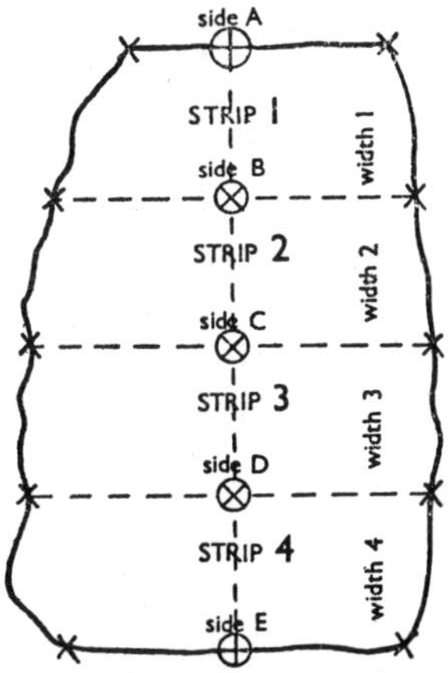

Fig. 53. Plot 2.

measured by means of a tape. These results are given in Table 25. Each strip is in fact a trapezium, so the area can be measured by the formula already given.

Total area of 955 square metres, converted to hectares, is 0·955 hectares.

TABLE 25

Results of the Survey of the Plot shown in Fig. 53

	Side 1 (metres)	*Side 2 (metres)*	*Height (metres)*	*Area (square metres)*
Strip 1	17	24	10	$\frac{(17+24) \times 10}{2} = 205$
Strip 2	24	26	10	$\frac{(24+26) \times 10}{2} = 250$
Strip 3	26	26	10	$\frac{(26+26) \times 10}{2} = 260$
Strip 4	26	22	10	$\frac{(26+22) \times 10}{2} = 240$
				TOTAL AREA 955

GUIDE TO FURTHER READING

Books

Chapter 1 to 5: Principles and Productive Resources

COHEN, R. L. (1949) *The Economics of Agriculture*, London, Nisbet, *Cambridge Economic Handbooks.*

DEXTER, K. and BARBER, D. (1961) *Farming for Profits*, Harmondsworth, Penguin Books.

HEADY, E. O. (1952) *Economics of Agricultural Production and Resource Use*, New York, Prentice-Hall.

HEADY, E. O. and JENSEN, H. (1957) *Farm Management Economics*, New York, Prentice-Hall.

INTERNATIONAL LABOUR OFFICE (1950) *Labour problems in Agriculture*, Geneva, I.L.O.

MARTIN, A. (1958) *Economics and Agriculture*, London, Routledge and Kegan Paul.

SCHULTZ, T. W. (1953) *Economic Organization of Agriculture*, New York, McGraw-Hill.

TUCK, R. H. (1961) *An Introduction to the Principles of Agricultural Economics*, London, Longmans.

YANG, W. Y. (1958) *Methods of Farm Management Investigations*, Rome, F.A.O., *Development Paper* 64,

Chapter 6: Soils and Crops

WRIGLEY, G. (1961) *Tropical Agriculture*, London, Batsford.

Chapter 7: Livestock

WILLIAMSON, G. and PAYNE, W. J. A. (1959) *Animal Husbandry in the Tropics*, London, Longmans.

Chapters 8 and 9: Records and Accounts

HOPKINS, J. A. and HEADY, E. O. (1951) *Farm Records*, Iowa State College Press.

JEFFREY, A. G. (1958) *Records and Accounts for Farm Management*, London, HMSO, *Ministry of Agriculture Bulletin* 176.

STURROCK, F. G. (1962) *Farm Accounting and Management*, London, Pitman.

Chapter 10: Farm Planning and Budgeting

BLAGBURN, C. H. (1961) *Farm Planning and Management*, London, Longmans.

MINISTRY OF AGRICULTURE, FISHERIES AND FOOD, London (1963) *The Farm as a Business*, London, HMSO.

PHILLIPS, T. A. (1964) *An Agricultural Notebook*, 2nd. ed., London, Longmans.

Appendix 1: Co-operatives

DIGBY, M. (1962) *Co-operatives*, London, Overseas Development Institute

SURRIDGE, B. J. and DIGBY, M. (1953) *A Manual of Co-operative Law and Practice*, Cambridge, Ordinances and Rules, Colonial Number 199.

Appendix 2: Surveying

DEBENHAM, F. (1961) *Map Making*, 3rd ed., Glasgow, Blackie and Son.

STEWART, F. R. and GRASSIE, J. C. (1947) *Surveying for Agritural Students and Planters*, London, George Allen and Unwin.

Journals

Farm Economist, Oxford.
Journal of Agricultural Economics, Edinburgh.
Journal of Farm Economics, Pennsylvania.

GLOSSARY

The meanings of some words and phrases found in this book.

accounts: Financial records of receipts and expenses.

accounting period: A period at the end of which accounts are sum marized.

bonus: An extra payment to workers above their normal wages.

budget: An estimate of the receipts and expenses of a proposed pla

business: A profit-making organization.

calving interval: The time interval between the birth of one calf an the birth of the next in the breeding history of a cow.

capital: The live- and deadstock and money necessary for carryin on a business.

cash crop: A crop which is normally sold rather than used at home o fed to livestock.

casual labour: Labour hired on a temporary basis.

communal labour: The members of the community working togethe on a particular project.

compost: Decayed plant material used as a manure.

conversion ratio: The weight of food fed to livestock divided by th liveweight gain.

co-operation: Operating together for mutual benefit.

crop processing: The treatment of crop products to increase thei market value.

crop rotation: A sequence of different crops grown in a regular order

deadstock: The farmer's stock (capital) other than growing crops an animals, i.e. buildings, machinery, stored products.

depreciation: The loss in value of capital items due to age and wea and tear.

diminishing returns (law of): If the rate of use of one productive re source is increased by successive equal steps, then the successiv amounts by which production is increased will eventually star to fall.

enterprise: A section or department of a farm.

fallow: A resting period to permit the natural build-up of fertility i the soil.

fixed costs: Costs which do not vary with the size of an enterpris and cannot be avoided by discontinuing production.

gross margin: The difference between the value of total production and variable costs.

humus: Dead and decaying plant and animal residues in the soil.

inoculation: Injection of a medicine through the skin of an animal to prevent disease.

intensive livestock: Livestock which can be kept on a very small area of land (e.g. pigs and poultry) and which, therefore, yield high returns per hectare.

interest: A payment made in return for the use of borrowed capital.

inventory: A detailed list of stock.

irrigation: The application of water to land by artificial means.

labour efficiency: The gross margin produced per man-hour of work.

lactation: The period during which an animal is producing milk.

land consolidation: The allocation of land so that each farmer can exchange his many scattered plots for a single holding.

livestock: Living animals, including poultry, kept by a farmer or stock-owner. In connection with valuations the term also includes growing crops.

man-hour: The work of one man for one hour.

market price: Local sale value.

mixed farming: The production of crops and animals together on the same farm.

mortgage: A loan obtained by offering land or buildings as security.

multiple cropping: Growing more than one crop per year on a piece of land, either by the planting of one crop through another (inter-cropping) or by the planting of one crop after another.

opportunity cost: The value of the return which would have been obtained if a productive resource had been employed in the best alternative way.

piece-work: Work which is paid for according to the amount done rather than by the time worked.

productive resources: The factors necessary for production: land, labour, capital and managerial ability.

profit: The gain from business activity. The excess of total production over costs.

profit and loss account: An account of the results of business activity over a period of time showing the profit or loss.

rent: A payment made in return for the use of land and buildings.

security for a loan: Anything offered as a pledge for repayment of a loan, to be forfeited in the case of non-payment.

sharecropping: A system whereby the land or crop is owned by one person and cultivated by another, the crop yields being shared between the two in a fixed proportion.

shifting cultivation: A system of land use in which a farmer clears an area of bush, cultivates crops in it for a few years and then abandons it to natural regrowth, and moves on to clear and cultivate other areas of bush.

soil conservation: The protection of the soil against erosion.

standard: A measure serving as a basis for comparison.

subsistence farming: Farming to produce just enough food to live on.

supplementary enterprise: An enterprise which may be added to existing enterprises without increasing the requirements of certain resources and without increasing all fixed costs.

tree crop: Crop which grows on trees. Perennial crops such as tea or coffee, which really grow on bushes, are often included in this term.

valuation: A statement of the value of the live- and dead-stock on a farm.

variable costs: Costs which vary with the size of an enterprise and can be avoided by discontinuing production.

work-peak: A large amount of work which must be done in a limited period, usually at a particular time of the year.

work study: The study of working methods (method study) and the time taken to complete individual operations (time study).

Index

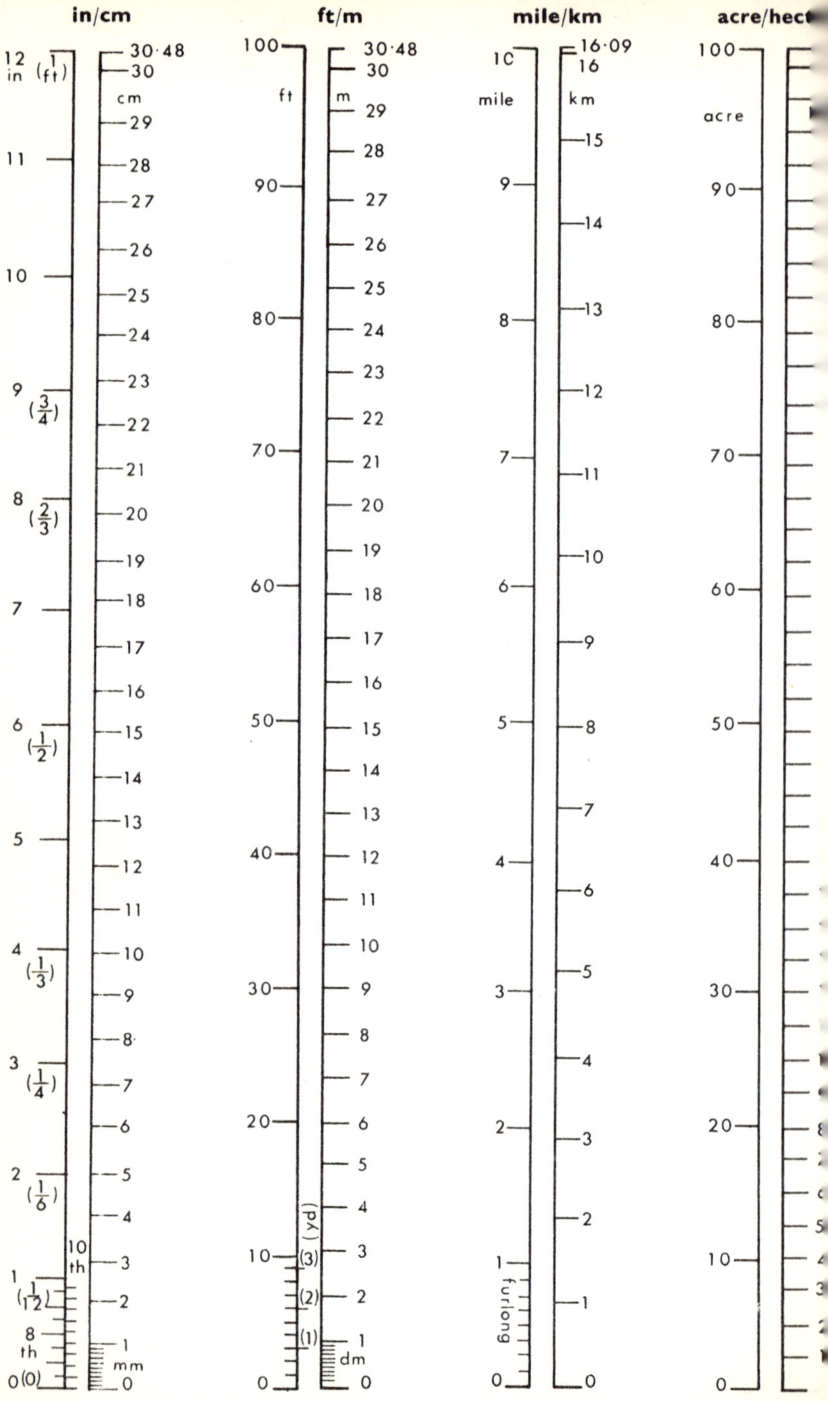
in/cm
ft/m
mile/km
acre/hect
in
(1/12 ft)
cm
30·48
(3/4)
(2/3)
(1/2)
(1/3)
(1/4)
(1/6)
(1/12)
10 th
8 th
mm
ft
m
(yd)
(3)
(2)
(1)
dm
mile
km
16·09
furlong
acre

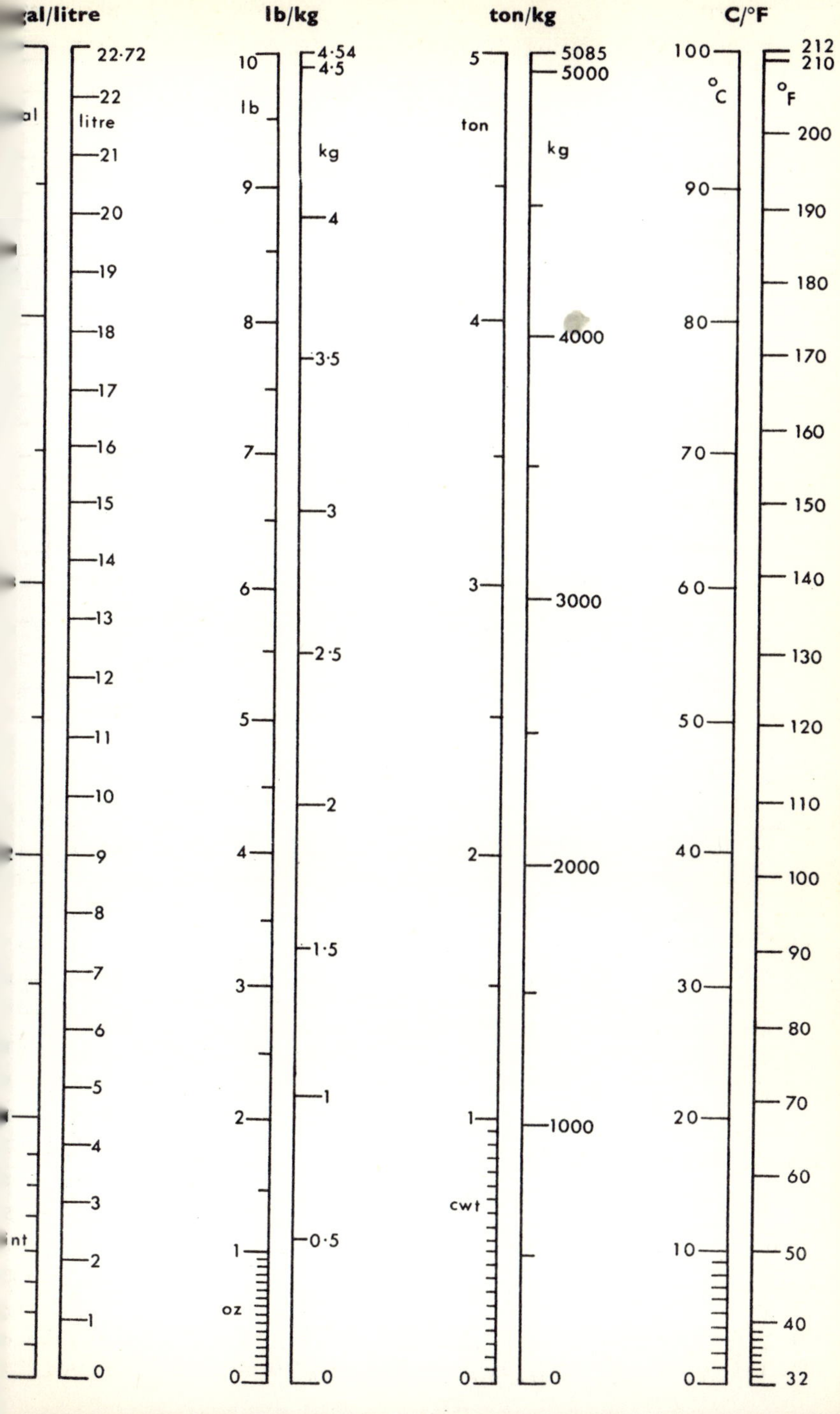
gal/litre
lb/kg
ton/kg
C/°F
al
int
litre
22·72
22
21
20
19
18
17
16
15
14
13
12
11
10
9
8
7
6
5
4
3
2
1
0
lb
oz
10
9
8
7
6
5
4
3
2
1
0
kg
4·54
4·5
4
3·5
3
2·5
2
1·5
1
0·5
0
ton
cwt
5
4
3
2
1
0
kg
5085
5000
4000
3000
2000
1000
0
°C
100
90
80
70
60
50
40
30
20
10
0
°F
212
210
200
190
180
170
160
150
140
130
120
110
100
90
80
70
60
50
40
32